Aircraft icing
A challenging problem of fluid mechanics.

Aircraft icing
A challenging problem of fluid mechanics.

Von der Fakultät für Maschinenbau
der Technischen Universität Carolo-Wilhelmina zu Braunschweig

angenommene Habilitationsschrift
zur Erlangung der Lehrbefugnis *(venia legendi)*
für das Lehrgebiet „Strömungsmechanik"

von
Dr.-Ing. Stephan E. Bansmer
aus Leipzig.

Braunschweig
5. September 2019

Bibliographical information held by the German National Library
The German National Library has listed this book in the Deutsche Nationalbibliografie (German national bibliography); detailed bibliographic information is available online at http://dnb.d-nb.de.
1st edition - Göttingen: Cuvillier, 2020

Nonnenstieg 8, 37075 Göttingen, Germany
Telephone: +49 (0)551-54724-0
Telefax: +49 (0)551-54724-21
www.cuvillier.de

1st edition, 2020
This publication is printed on acid-free paper.

ISBN 978-3-7369-7224-7
eISBN 978-3-7369-6224-8

Page layout inspired by the work of Edward R. Tufte and Donald E. Knuth. `tufte-latex.googlecode.com`
The watercolour painting on page 3 is inspired by the artistic work of Evelyn Gerbaud.

The presented work was funded by the following projects, institutions and grants: DLR@Uni SuLaDI (Helmholtz Association), the European Union's Seventh Framework Programme in Research, Technological Development and Demonstration (grant agreement n° ACP2-GA-2012-314314), the German Federal Ministry of Economic Affairs and Energy (grant number Drífa-FKZ0325842A), and the German Research Foundation (DFG grants BA4953/2-1, BA4953/3-1, BA4953/3-2, BA4953/5-1). The major subcontractors involved in building the icing wind tunnel were HST GmbH, GWA Gesellschaft für Wärme- und Anlagentechnik mbH, and Greencooltec GmbH. The author further expresses his thanks to all project partners from Europe and overseas.

The manuscript was edited for English language usage, grammar, spelling and punctuation by editors at the Nature Research Editing Service and at American Journal Experts.

First printing, June 2020

to M and l_1

"Why is it that when you awake to the world of realities you nearly always feel, sometimes very vividly, that the vanished dream has carried with it some enigma which you have failed to solve? You smile at [...] your dream, and yet you feel that this tissue of absurdity contained some real idea, something that belongs to your true life, – something that exists, and has always existed, in your heart."

— Fyodor M. Dostoyevsky (1869),
translated by Eva Martin (1915)

Contents

Outline

In 2017, approximately 4.1 billion passengers were carried by scheduled flights in the member states of the International Civil Aviation Organization (ICAO). To satisfy the needs of day and night flight operations even in situations of inclement weather, aircraft icing is an inherent component of the certification of large transport aircraft. Ice detection and protection systems ensure flight safety and must be designed carefully. The reliable prediction of ice accretion is a mandatory intermediate step that involves many challenging problems of *fluid mechanics* that will be treated in this volume.

THIS MANUSCRIPT is divided into three main parts. Part I introduces the topic of aircraft icing with a brief historical overview of technological and scientific progress over the last century. Thereafter, challenges from the perspective of fluid mechanics are introduced. In particular, the governing boundary conditions of the phenomenon are highlighted, accompanied by remarks on non-dimensional groups and scaling laws. Finally, selected physical properties of water and ice are elucidated.

PART II emphasizes small-scale phenomena. Classic ice accretion is usually triggered by the impact of supercooled droplets on a surface. After analysing the solidification process of supercooled water, the goal is to study the drop impact on both dry and wetted substrates.

Figure 1: First experiment in the Braunschweig Icing Wind Tunnel, August 2013. Typical features of ice accretion on a NACA0012 aerofoil are visible.

Indeed, the surface properties strongly affect the impact dynamics and hence the entire course of icing.

The macroscopic phenomena of ice accretion are considered in Part III using a combined approach of experiments and computations. To carry out experimental investigations, the Braunschweig Icing Wind Tunnel was built, see Figure 1. Together with the results of the icing code TAUICE, the gathered data are used to develop a deeper understanding of the process of glaze ice accretion, which involves a broad range of physical phenomena, including wetting, roughness formation, transitional flow and heat transfer.

Acknowledgements

First of all, I would like to thank Rolf Radespiel, the supervisor of my *Habilitation* in Fluid Mechanics, who made it possible for me to establish a working group on multi-phase flows and icing at the Institute of Fluid Mechanics at the TU Braunschweig in 2011. I would like to thank him for this leap of faith and the great degree of freedom he granted me during my research work.

The scientific knowledge gained since then forms the basis of this habilitation thesis. Therefore, I would especially like to thank all members of my working group: Arne Baumert, Venkatesh Bora, Ana Carolina Botelho, David Burzynski, Oliver Esselmann, Xiaoliang Gong, Charlotte Hertel, Inken Knop, Vivien Muxfeldt, Stephan Sattler, Denis Khavier Sotomayor Zakharov, and Juan Velandia.

My gratitude also goes to all members of the Institute of Fluid Mechanics and the Faculty of Mechanical Engineering of the TU Braunschweig, who supported and accompanied me during my tenure. In particular, I would like to mention Rinie Akkermans, Markus Böl, Günter Bräuer, Andreas Dietzel, Jens Friedrichs, Georg Garnweitner, Peter Hecker, Christoph Herrmann, Peter Horst, Ulrike Krewer, Stephan Scholl, Peter Scholz, Richard Semaan, Michael Sinapius, and Enrico Stoll.

I would like to thank Éric Laurendau and Christian Kähler for the review of my habilitation thesis. I owe the latter my passion for flow measurement technology.

Since the establishment of the working group, I could fully count on the support of the German Aerospace Center (DLR) at any time. In particular, I would like to thank Heinz Voggenreiter, Cord Rossow, Stefan Levedag, Christiane Voigt, Norbert Kroll, Olaf Brodersen, Jan Delfs, Bernhard Eisfeld, Stefan Görtz, Tina Jurkat-Witschas, Per Ohme, Axel Probst, and Jan Steiner.

I was involved with many projects in an international research framework and was able to benefit from the exchange with leading scientists and development engineers. I would like to give special thanks to André Altmikus, Christian Bartels, Tom Currie, Fabien Dezitter, Biagio Esposito, Dan Fuleki, Franck Hervy, Harry Hoeijmakers, Fassi Kafyeke, Aleksey Miller, Ilia Roisman, Alfons Schwarzenboeck, Cameron Tropea, Paolo Vanacore, and Philippe Villedieu. Moreover, I would like to thank Nikolaus Adams, Holger Babinsky, and Wagdi Habashi for allowing me to spend several months researching at their universities.

Finally, I would like to sincerely thank my family. They are my haven of peace and have offered me support in all situations of life.

Part I

General introduction

1
Historical remarks and current airframer perspectives on icing

With the growing need for scheduled flights under all weather conditions since the 1920s, the subject of aircraft icing involves nearly one century of research, technology development and cross-fertilization over different engineering disciplines. Without being exhaustive, the timeline on page 15 identifies major milestones and knowledge drivers as well as accidents and failures that form the current view of aircraft icing. This chapter summarizes these developments and provides information on current certification aspects for flight in icing conditions.

1.1 Milestones in icing research

The 1920s and 1930s can be considered the dawn of modern aviation. In aircraft design, anodized aluminium replaced wooden structures. Pioneers like Albert Read, Charles Lindbergh and Charles Kingsford Smith crossed the oceans. Carriers like *Koninklijke Luchtvaart Maatschappij (KLM)* and *Deutsche Luft Hansa* were founded. The demand for scheduled day-and-night flight operations grew rapidly. Soon, aircraft icing was recognized as a great obstacle. To circumvent ice accretion on the leading edge of the wings, pneumatic boots were developed for de-icing; a Northrop Alpha mail plane was the first commercial aircraft equipped with rubber blankets from Goodrich. A good summary of the earliest research on aircraft icing can be found in the Russian literature.[1]

[1] Nikolai V. Lebedev. *Combating ice on aircraft*. State publishing house of defense industry, People's Commissariat of Aviation Industry, 1939

During World War II, much theoretical work on ice accretion was performed by the Nobel laureate Irving Langmuir. Perhaps stimulated by his experience as a private pilot but also motivated by the US Army Air Forces, Langmuir studied electrostatic precipitation on aircraft flying through snowstorms, the supercooling of droplets and ice adhesion on metal surfaces.[2] By combining his mathematical theory on droplet trajectories with data on cylinder ice accretion gathered on Mount Washington, he eventually developed a pioneer-

[2] Irving Langmuir. *The collected works of Irving Langmuir*, volume 10. Pergamon Press, 1961

ing method for establishing cloud droplet sizes and their liquid water content.

THE DEBUT of the de Havilland Comet in 1952 ushered in a new era of jet airliners. Thermal ice protection systems using the bleed air of jet engines became standard. However, enormous heat requirements[3] of more than 10 percent bleed air were predicted, precluding continuous heating operation. It was Bernhard Messinger from Lockheed Aircraft Corporation who suggested in his seminal paper[4] in 1953 the design of cyclic thermal de-icing systems based on a thorough analysis of the heat and mass balances of ice accretion. However, the first studies were limited to simple geometries, such as cylinders, for which analytical solutions were deducible. The technology development thus still relied on rig tests in icing wind tunnels. The major icing wind tunnels of that time were NASA's Icing Research Tunnel[5] and the NRC's Low Speed Icing Tunnel.

[3] Thomas F. Gelder, James P. Lewis, and Stanley L. Koutz. Icing Protection for a Turbojet Transport Airplane: Heating Requirements, Methods of Protection, and Performance Penalties. Technical Report NACA-TN-2866, National Advisory Committee for Aeronautics, 1953

[4] Bernard L. Messinger. Equilibrium temperature of an unheated icing surface as a function of air speed. *Journal of the Aeronautical Sciences*, 20(1): 29–42, 1953

[5] William M. Leary. "We Freeze to Please": A History of NASA's Icing Research Tunnel and the Quest for Flight Safety. Technical Report NASA/SP-2002-4226, National Aeronautics and Space Administration, 2002

INTEGRATED CIRCUITS in the 1960s and microprocessors like Intel's 4004, which was commercially launched in 1971, enabled breakthroughs in the vast and growing fields of electronics and computers. Figure 1.1 shows the exponential increase in computational power over the period of seventy years in which the field of computational fluid dynamics evolved from solvers of linear potential equations to Euler equations in the 1980s. During that time, cooperation among NASA, RAE and ONERA resulted in the first generation of icing codes[6] that incorporated much of the earlier theoretical work of LANGMUIR and MESSINGER. NASA's LEWICE code is one of these ice prediction tools that remains the standard in many industrial fields. After the end of the Cold War, many other agencies joined icing research activities, and a vast number of icing codes emerged that took advantage of solving the Reynolds-averaged Navier-Stokes equations for improved boundary layer predictions. The AGARD advisory report 344 summarizes the status of ice accretion simulation as of 1997.[7]

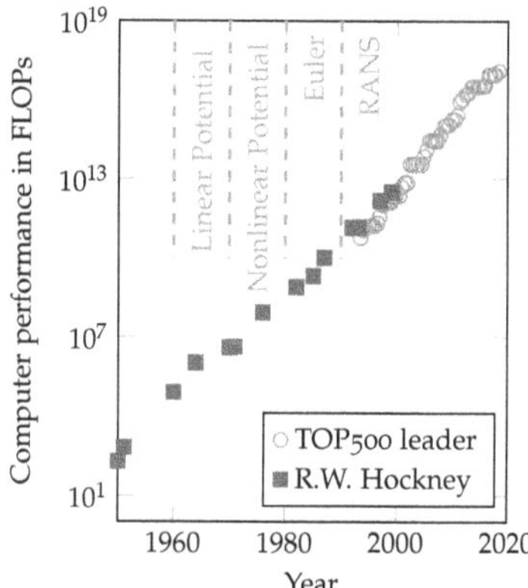

Figure 1.1: Development of computational performance over the last 70 years. Chronology of computational methods for fluid dynamics in ANTONY JAMESON's view.

PARALLEL TO THIS COMPUTATIONAL PROGRESS, developments in LASER and digital image recording enriched experimental icing research. The introduction of phase Doppler interferometry for particle sizing in 1984 laid the foundations for more accurate characterizations of atmospheric icing conditions and icing wind tunnel experiments. As a consequence, details of the icing process came into focus in experimental studies. The visualization of ice roughness growth by OLSEN and WALKER (1986) and investigations on water film dynamics during ice accretion by MYERS and HAMMOND (1999) are just a

[6] Roger W. Gent, Nicholas P. Dart, and James T. Cansdale. Aircraft icing. *Philosophical Transactions of the Royal Society of London A: Mathematical, Physical and Engineering Sciences*, 358 (1776):2873–2911, 2000

[7] Bernhard Wagner. Ice Accretion Simulation. Technical Report AGARD report no. 344, North Atlantic Treaty Organization, 1997

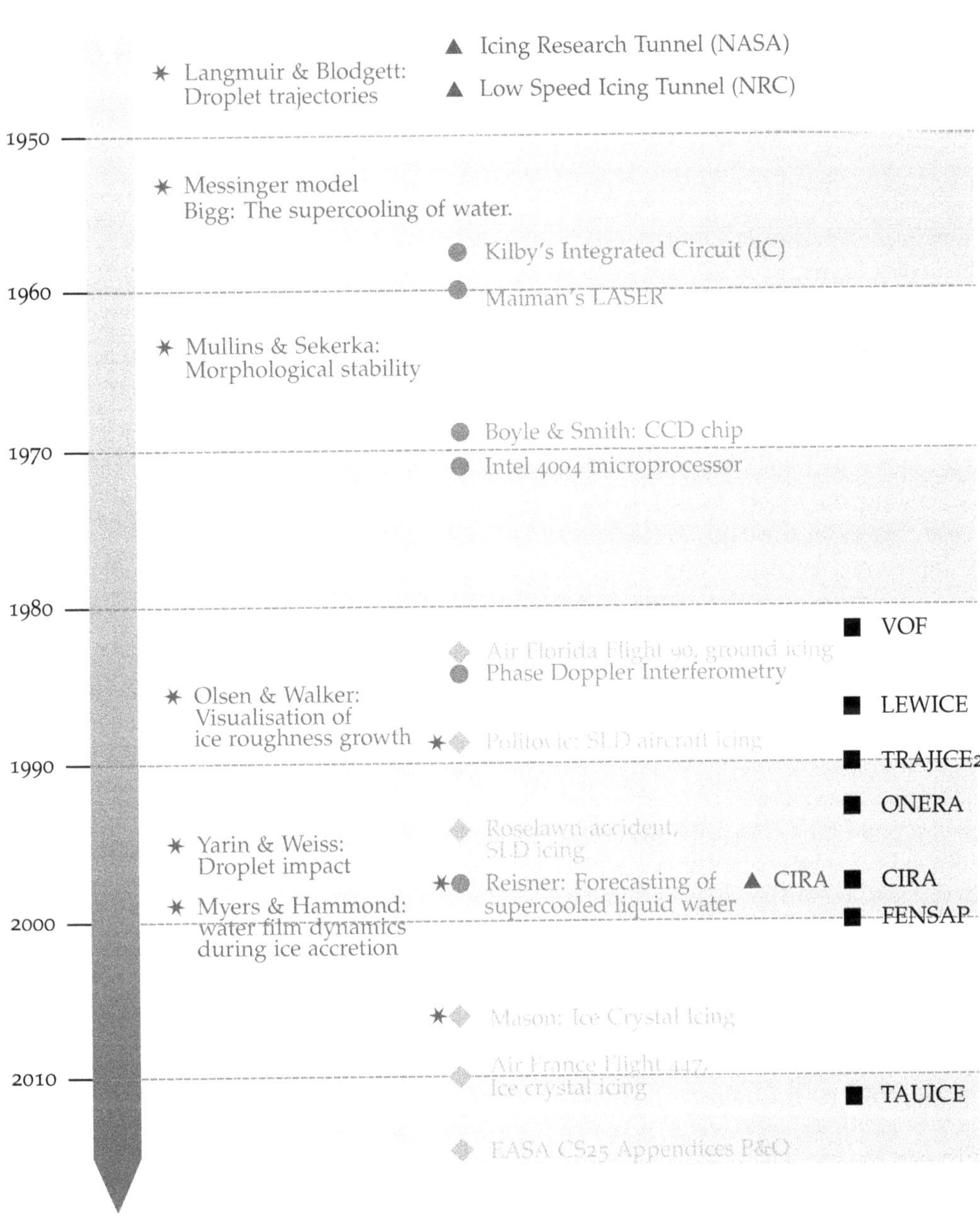

Figure 1.2: A glimpse of scientific publications, technology enablers, drivers and developments that shaped the field of modern icing research.

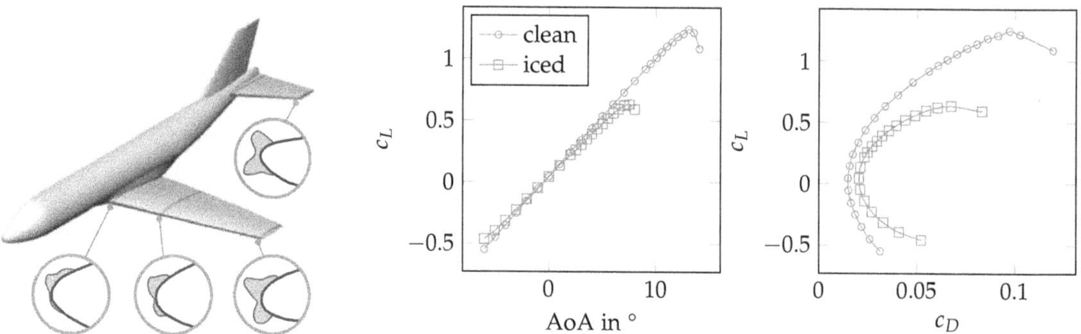

Figure 1.3: Half-model of the VFW 614 ATTAS aircraft including simulated leading edge ice accretion on the main wings and horizontal tail plane. The aerodynamic performance is significantly deteriorated compared with that of the clean configuration. Adapted picture and data from DEILER and KILIAN (2018).

few examples. The new findings improved the underlying computational models, although further improvements are an ongoing effort in the aircraft icing community.

1.2 *Certification of commercial transport aircraft*

Before entering into operation, a newly developed aircraft model must obtain a type certificate issued by the responsible aviation regulatory authority. These agencies include the Civil Aviation Directorate in Canada, the Federal Aviation Administration (FAA) in the US, and the European Aviation Safety Agency (EASA) in Europe. Depending on the aircraft type, specific guidelines for certification are applicable to demonstrate compliance with the regulatory requirements. For large transport aircraft in Europe, the EASA CS-25 document lists the certification specifications. Articles CS25.1419, CS25.1420 and CS25.1093 and appendices C, O and P are relevant for certification for flight in icing conditions. Three major icing phenomena are addressed: supercooled droplet icing, supercooled *large* droplet icing[8] and mixed phase icing.[9] The latter phenomenon was discovered only recently and encompasses ice crystals, which can become sticky on warm surfaces, like in Pitot probes and in the low-pressure compressor stages of jet engines.

[8] Marcia K. Politovich. Aircraft Icing Caused by Large Supercooled Droplets. *Journal of Applied Meteorology*, 28(9): 856–868, 1989

[9] Arne Baumert, Stephan E. Bansmer, Pierre Trontin, and Philippe Villedieu. Experimental and numerical investigations on aircraft icing at mixed phase conditions. *International Journal of Heat and Mass Transfer*, 123:957–978, 2018

BUILDING ON THE PREDICATIVE CAPABILITY of modern icing codes, the philosophy of icing certification has shifted in the last two decades from complete reliance on rig testing towards design by computational analysis. Figure 1.3 highlights the possibilities of state-of-the-art tools applied to a VFW 614 ATTAS aircraft model. In the first step, ice accretion on the wings and the horizontal tail plane is simulated using the TAUICE code. Note the different ice shapes along the wingspan, which are called glaze ice horns. The deteriorating aerodynamic performance of the configuration is evaluated with

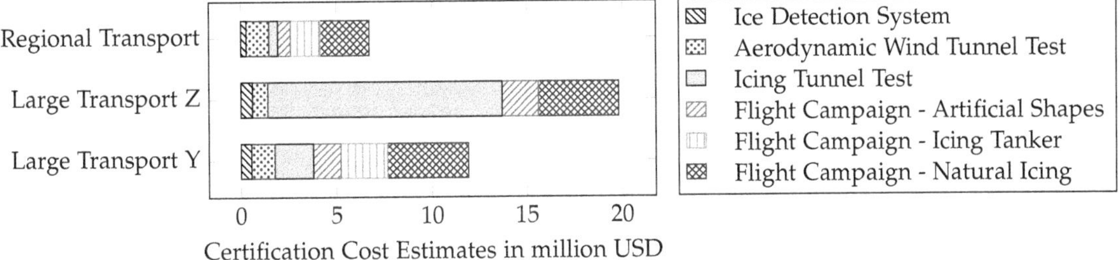

Figure 1.4: EASA Part 25 Appendix O certification cost estimates. Data source: Working group report on supercooled large droplet rulemaking.

the DLR flow solver TAU. When compared with clean conditions, the maximum lift decreases, and both the stall speed and aerodynamic drag increase. As a consequence, the flight envelope of the aircraft is significantly reduced, which poses a serious threat to flight safety, especially during the flight phase of holding and approach to landing. Details of the computational approach will be discussed in Chapter 9.

DESPITE COMPUTATIONAL ADVANCES, wind tunnel and flight tests are still an important pillar in the certification procedure when validation data are sparse. One example is the new requirements for compliance with supercooled large droplet (SLD) icing that were issued only recently in 2016. Due to their increased inertia, large droplets can impact and accrete further downstream in regions that might not be protected by classical anti-icing systems. The study of ice accretion is therefore always linked to the design task of ice protection systems. The power requirements of an anti-icing system are usually approximately 150 kW, regardless of the aircraft size; for small aircraft, nearly the entire leading edge must be protected, while large aircraft meet the requirements by protecting a limited number of slats. Cost estimates for the SLD-icing certification procedures from different airframers are consolidated in Figure 1.4.

FLIGHT SAFETY is the driving paradigm in the entire certification process. Continued progress in the fields of computational analysis and experimental testing procedures maintains high safety standards. Statistics from NASA's Aviation Safety Reporting System (ASRS) and the Bureau of Transportation Statistics (BTS) emphasize that trend, as shown in Figure 1.5. While the number of enplanements has steadily grown over recent decades, the numbers of incidents and accidents remain rather small.

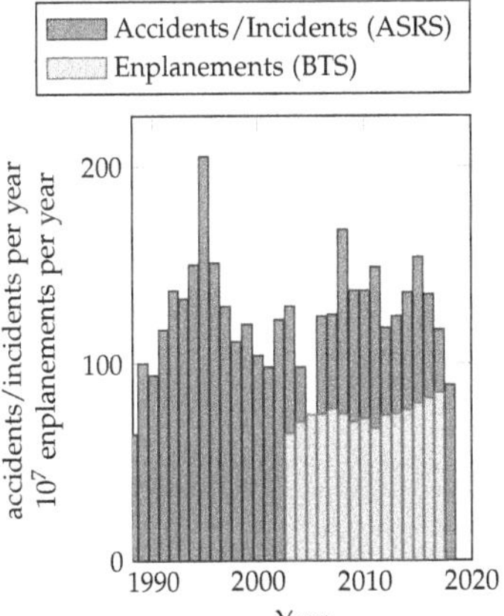

Figure 1.5: Statistics on icing incidents/accidents in the US airspace. Data sources: ASRS and BTS. Data search methodology adapted from GREEN (2006).

2
A challenging problem of fluid mechanics

Supercooled droplet icing is the most frequently encountered scenario for aircraft icing. It involves fluid mechanics on various time and length scales and covers fundamental problems like drop impact, turbulent flow, and non-equilibrium thermodynamics. Consequently, the definition and assessment of many boundary conditions are mandatory.

2.1 *Supercooled droplet icing: phenomenology and scales*

A droplet is called supercooled when it is in liquid phase but its temperature is below the freezing point. The lifetime of this metastable state varies from a few seconds to several days, depending on the droplet's size and the difference in temperature from the equilibrium freezing point at 0 °C.[1] Minor perturbations can initiate the phase change. Hence, the drop solidifies upon impact onto a solid substrate, which is the basis of ice accretion.

[1] Hans R. Pruppacher. A new look at homogeneous ice nucleation in supercooled water drops. *Journal of the Atmospheric Sciences*, 52(11):1924–1933, 1995

The temporal evolution of ice accretion on an NACA0012 aerofoil is visualized in Figure 2.1. At $t_{acc} = 0$, the initially dry aluminium aerofoil is exposed to a cloud of supercooled droplets. After five seconds, the surface is fully wetted in the vicinity of the stagnation line. A thin water film forms, whose liquid/gas interface is subject to several instability mechanisms, including the formation of

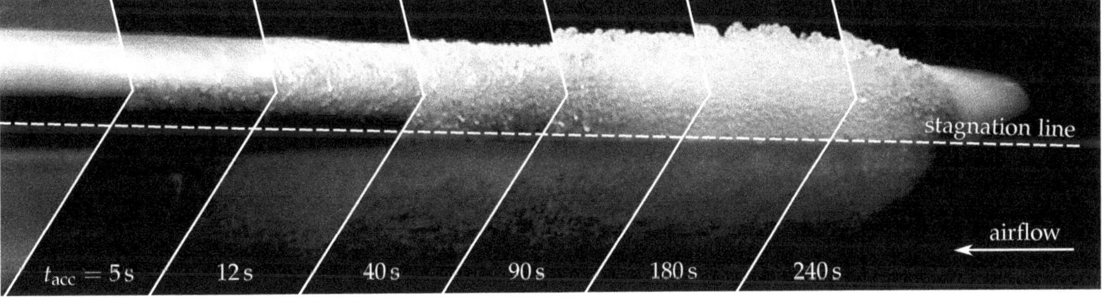

Figure 2.1: Ice accretion on an NACA0012 aerofoil over time. $T_\infty = 263.15$ K, LWC $= 1.3\,\mathrm{g\,m^{-3}}$, MVD $= 80\,\mu$m, AoA $= 8°$. Data from Sommerwerk, Horst and Bansmer (2016).

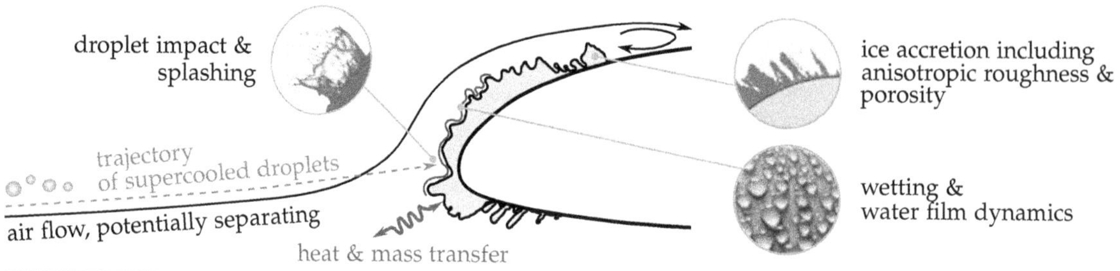

Figure 2.2: Summary of fluid mechanical phenomena associated with aircraft icing.

rivulets. Subsequently, solidification commences, and small roughness elements are formed that grow over time. Two minutes later, the accreted ice layer already has a thickness of several centimetres.

Obviously, several fluid mechanical phenomena are involved in the icing process, which are summarized in Figure 2.2. Each of these individual phenomena has attracted many researchers in the past, as reflected by the cited ground-breaking literature and Nobel laureates. The major challenges can be outlined as follows:

- **Droplet trajectories.** The gravity, aerodynamic drag, and inertia of the droplet imply trajectories[2] that deviate from the airflow streamlines. These trajectories govern the mass balance of ice accretion.

- **Droplet impact** at high velocities promotes splashing.[3] Secondary droplets are generated, which alter the local mass balance of ice accretion and might re-impinge on installations further downstream.

- **Large ice roughness**, partially wetted by thin water films.[4] Anisotropic roughness elements are observed that severely interfere with the aerodynamic boundary layer. The roughness height can vary by about one order of magnitude in the streamwise direction.

- **The laminar/turbulent flow transition** is shifted upstream due to abundant roughness elements that create large-amplitude disturbances in the region of laminar flow.[5]

- **Heat and mass transfer.** A broad range of thermodynamic processes[6] are involved in aircraft icing, particularly kinetic and viscous heating, convection, conduction, solidification and evaporation. Since the thermal and concentration boundary layers exhibit similarities to the velocity boundary layer, the aforementioned challenges of roughness and water film dynamics fully apply to the problem of heat and mass transfer.

[2] Irving Langmuir and Katherine B. Blodgett. *The collected works of Irving Langmuir*, volume 10 - Atmospheric phenomena, chapter Mathematical investigation of water droplet trajectories, pages 335–393. Pergamon press, 1961

[3] Alexander L. Yarin. Drop impact dynamics: splashing, spreading, receding, bouncing... *Annual Review of Fluid Mechanics*, 38:159–192, 2006

[4] Pyotr L. Kapitza and Sergey P. Kapitza. Wave flow of thin viscous fluid layers. *Zh. Eksp. Teor. Fiz*, 18(1): 3–28, 1948

[5] Hugh L. Dryden. Review of published data on the effect of roughness on transition from laminar to turbulent flow. *Journal of the Aeronautical Sciences*, 20(7):477–482, 1953

[6] Ilya Prigogine. *Introduction to thermodynamics of irreversible processes*. Interscience Publ., 3. ed. edition, 1967. ISBN 0470699280

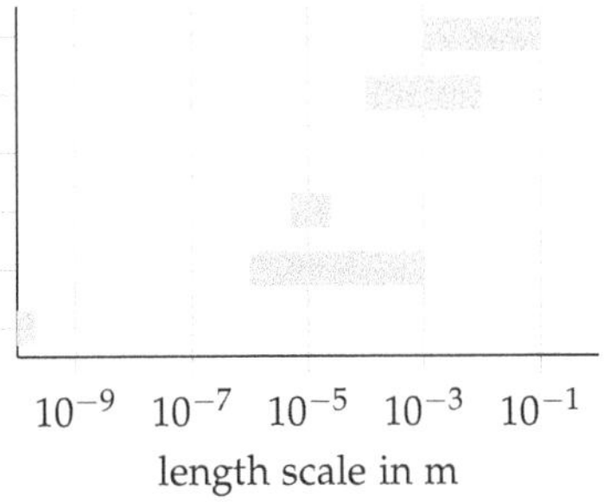

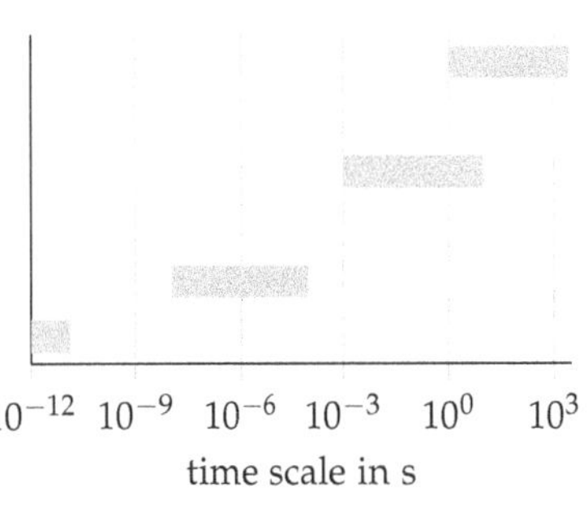

Figure 2.3: A selection of length and time scales involved in aircraft icing.

- **Flow separation.** Once a sufficient amount of ice is accumulated, the airflow might separate,[7] forming a recirculation zone behind the accretion. Simulating such detached flows is still a computational challenge.[8]

[7] Roger L. Simpson. Turbulent Boundary-Layer Separation. *Annual Review of Fluid Mechanics*, 21(1):205–232, 1989

[8] Philippe R. Spalart. Detached-eddy simulation. *Annual Review of Fluid Mechanics*, 41:181–202, 2009

SEVERAL LENGTH AND TIME SCALES form the basis of ice accretion, as shown in Figure 2.3. These scales range from molecular reorientation within picoseconds during phase change to macroscopic ice growth, which can last several hours or longer. The present monograph is therefore divided into another two parts that consider the micro- and macro-scales of ice accretion separately.

2.2 *Boundary conditions and scaling laws*

The multitude of phenomena necessitate the definition of several boundary conditions. Most importantly, the airflow and droplet cloud must be characterized.

Boundary conditions for airflow

Fluid motion is governed by the conservation of mass, momentum and energy. Mathematically, this concept can be described with transport equations that are further outlined on page 86. Their analysis reveals three major physical effects relevant to icing that must be characterized with appropriate boundary conditions:

- **Inertial effects.** A body $\mathscr{B}$ moves with velocity U_∞ through a fluid with density ϱ_{air}. The body locally deflects the flow, yielding inertial forces according to NEWTON's second law.
- **Viscous effects.** The dynamic viscosity of air μ_{air} will mainly influence the flow inside a thin region near the surface of the body, which is called the *boundary layer*.[9]

[9] Hermann Schlichting. *Boundary-Layer theory*. Springer, Berlin, 9th edition, 2017. ISBN 9783662529171

LWC fraction *in* %	5	10	20	30	20	10	5
$\left(\frac{d}{\mathrm{MVD}}\right)^2$	0.31	0.52	0.71	1	1.37	1.74	2.22

Table 2.1: Langmuir-D distribution of droplet size.

- **Compressibility.** When the static pressure p_{air} of air increases, its density ϱ_{air} will increase. Under isentropic conditions, the relation $a_{\mathrm{air}}^2 = \frac{\partial p_{\mathrm{air}}}{\partial \varrho_{\mathrm{air}}}$ applies, where a_{air} represents the speed of sound.[10] Compressibility manifests itself at higher flow velocities, where the pressure variations around the body become large compared with the absolute pressure.

[10] John David Anderson. *Modern compressible flow*. McGraw-Hill series in aeronautical and aerospace engineering. McGraw-Hill, 3rd edition, 2003. ISBN 0071121617

Boundary conditions for the droplet cloud

The airflow is laden with a cloud of supercooled droplets, which can be characterized by the following parameters:

- Water concentration in the air, which is quantified by the **liquid water content** (LWC), a measure of the mass of water per unit volume of air. Typical values for atmospheric icing conditions range from $0.1\,\mathrm{g\,m^{-3}}$ to $3\,\mathrm{g\,m^{-3}}$.

- **Median volume diameter** (MVD) of the statistical distribution of water droplets in the liquid cloud. The MVD attempts to reduce the size distribution to a single representative scalar diameter. This idea is attributed to LANGMUIR.[11] His theory of droplet trajectories relates droplet size distributions and ice accretion on a cylinder. After exposing a cylinder to a cloud of super-cooled fog, Langmuir derived an assumption of the droplet size spectra, which can be found in Table 2.1. The term "median" results from the median value of the LWC in this distribution, where $d/\mathrm{MVD} = 1$. Langmuir himself described the MVD as a diameter "such that half of the liquid water content of the fog is in droplets of larger [diameter] and the other half in droplets of smaller [diameter]."

[11] Irving Langmuir and Katherine B. Blodgett. *The collected works of Irving Langmuir*, volume 10 - Atmospheric phenomena, chapter Mathematical investigation of water droplet trajectories, pages 335–393. Pergamon press, 1961; and Karen J. Finstad, Edward P. Lozowski, and Lasse Makkonen. On the median volume diameter approximation for droplet collision efficiency. *Journal of the Atmospheric Sciences*, 45 (24):4008–4012, 1988

- **Humidity**, i.e., the amount of water vapour in the air, can influence the freezing process of supercooled droplets.[12] A distinction is made between the absolute humidity (AH) and relative humidity (RH). AH is defined by the mass of water vapour per unit volume of dry air. With the partial pressure of the water vapour p_{vap}, the static temperature T_{air} of the air and the specific gas constant of the vapour $\mathscr{R}_{\mathrm{vap}}$, this definition yields $\mathrm{AH} = \frac{p_{\mathrm{vap}}}{\mathscr{R}_{\mathrm{vap}} T_{\mathrm{air}}}$. RH is defined as the ratio of the partial pressure of water vapour p_{vap} to the equilibrium vapour pressure of water $p_{\mathrm{vap}}^{\bullet}$, i.e., $\mathrm{RH} = \frac{p_{\mathrm{vap}}}{p_{\mathrm{vap}}^{\bullet}}$.

[12] Stefan Jung, Manish K. Tiwari, N. Vuong Doan, and Dimos Poulikakos. Mechanism of supercooled droplet freezing on surfaces. *Nature Communications*, 3:615, January 2012

Rime and glaze ice

Ice accretion is initiated when a cloud of supercooled droplets collides with the body $\mathscr{B}$. The **static air temperature** T_{air} governs the subsequent solidification process. At very low temperatures, the droplets will freeze shortly after their impingement and entrap the surrounding air. The resulting ice accretion is called rime ice; see Figure 2.4. As the temperature increases, the solidification process of the impacting droplets is retarded, promoting wind-driven water film dynamics. At locations with increased convective heat transfer, the water film freezes, yielding a nearly transparent glaze ice shape with typical horn formations. The list of boundary conditions is completed by the **accumulation time** t_{acc} for which the body $\mathscr{B}$ is exposed to the cloud of super-cooled droplets.

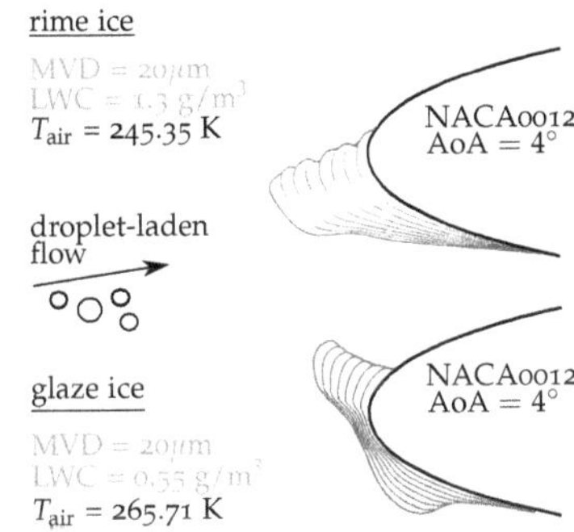

Figure 2.4: Rime ice and glaze ice shapes over time at the leading edge of an NACA0012 aerofoil, chord length 0.53 m. One blue line represents one minute of ice accretion. Further boundary conditions (rime): $U_\infty = 58\,\text{m}\,\text{s}^{-1}$, $t_{\text{acc}} = 8\,\text{min}$; (glaze): $U_\infty = 102.8\,\text{m}\,\text{s}^{-1}$, $t_{\text{acc}} = 7\,\text{min}$. Data from computations with TAUICE by Jan Steiner.

Non-dimensional groups and scaling laws

For airworthiness certification, aircraft and engine manufacturers perform many icing wind tunnel experiments with subscale models before full-scale flight tests. For such subscale tests, all the above-mentioned dimensional boundary conditions have to be adapted to produce an icing result similar to that obtained in a full-scale reference model. Furthermore, limited ranges of air speed, test section dimensions and icing cloud characteristics in the icing wind tunnels have to be contemplated. In the first step, non-dimensional groups must be deduced from the boundary conditions.[13]

[13] Thomas H. Bond and David N. Anderson. Manual of Scaling Methods. *NASA/CR-2004-212875*, 2004

AIRFLOW SIMILARITY requires an interrelation among inertial, viscous and compressible effects. For this purpose, both the Reynolds number and Mach number are introduced:

$$\text{Re}_{\text{air}} = \frac{\varrho_{\text{air}} \cdot U_\infty \cdot d_{\text{nose}}}{\mu_{\text{air}}} \qquad \text{Ma}_{\text{air}} = \frac{U_\infty}{a_{\text{air}}} \quad .$$

Because icing mainly occurs in the vicinity of the leading edge of an aerofoil, the reference length d_{nose} in the Reynolds number definition is chosen as twice the nose radius.

THE TRAJECTORIES of supercooled droplets provide information on if and where a droplet will impinge on the aerofoil surface. They are hence an important element for determining the mass balance of ice accretion. Analysing the equation of droplet motion – cf. Chapter 9.3 – suggests a non-dimensional inertia parameter $\mathfrak{K}$, given by

$$\mathfrak{K} = \frac{\varrho_{\text{d}} \cdot \text{MVD}^2 \cdot U_\infty}{18 d_{\text{nose}} \cdot \mu_{\text{air}}} \quad ,$$

Method	Similarity parameters							
	Re_{air}	Ma_{air}	$\mathfrak{K}$	A_c	f_0	Y_d	Y_{air}	Y_{rel}
Charpin and Fasso (1972)			*	*	*			*
Ruff (1986) – no. 1			*	*				
– no. 2			*	*	*			
– no. 3			*	*	*			*
– no. 4			*	*	*	*	*	*

Table 2.2: A selected number of icing similarity studies.

where ϱ_d denotes the droplet's density. The impinging mass flux of supercooled droplets is transformed into a non-dimensional accumulation parameter A_c:

$$A_c = \frac{\mathrm{LWC} \cdot U_\infty \cdot t_{acc}}{\varrho_{ice} \cdot d_{nose}} \quad .$$

Herein, ϱ_{ice} represents the density of ice.

THE ENERGY TRANSFER among air, impacting droplets, ice and water film involves many physical phenomena. See Chapter 9.4 for a detailed discussion. Many non-dimensional parameters can be derived, among which the stagnation line freezing fraction f_0 has become very popular:

$$f_0 = \frac{c_{p,ws}}{L_{fus}} \left[Y_d + \frac{Y_{air}}{Y_{rel}} \right] \quad ,$$

where

$$Y_d = 273.15\,\mathrm{K} - T_{\infty,air} - \frac{U_\infty^2}{2c_{p,ws}}$$

$$Y_{air} = T_S - T_{\infty,air} - \frac{U_\infty^2}{2c_{p,air}} + \frac{1}{c_{p,air}} \left(\frac{\mathrm{Pr}_{air}}{\mathrm{Sc}_{air}} \right)^{\frac{2}{3}} \frac{p_{ww} - p_w}{p_{\infty,air}} L_{evap}$$

$$Y_{rel} = \frac{d_{nose}}{\mathrm{Nu}_{air}\, k_{air}} \mathrm{LWC}\, U_\infty \beta_0 (\mathfrak{K})\, c_{p,ws} \quad .$$

The stagnation line freezing fraction can be interpreted as the percentage of water at the stagnation line that will freeze. Note that the parameter Y_{rel} is dimensionless, while Y_d and Y_{air} have units of temperature.

$c_{p,air}$	constant-pressure specific heat of air
$c_{p,ws}$	specific heat of water on model surface
k_{air}	thermal conductivity coefficient
L_{fus}	latent heat of fusion
L_{evap}	latent heat of vaporization
p_{ww}	vapour pressure of water over liquid water
p_w	vapour pressure of water in the atmosphere
Nu_{air}	Nusselt number
Pr_{air}	Prandtl number
Sc_{air}	Schmidt number
Y_{air}	air energy transfer parameter
Y_d	drop energy transfer parameter
Y_{rel}	relative heat factor
β_0	stagnation-line catch efficiency

THE ACCURATE SCALING of the test conditions between the reference and subscale model would require matching all the above non-dimensional groups. Merely matching the Reynolds and Mach numbers in classical aerodynamics is seldom possible. A rigorous application of all non-dimensional groups for scaling aircraft icing is therefore unfeasible. For practical considerations, the main goal for adequate scaling is to simulate the characteristic features, such as the glaze horns, in size, location, and shape. Based on systematic empirical studies, different simplifications have been introduced, e.g., by ignoring some of the non-dimensional groups; see Table 2.2. Apparently, the airflow similarity is not of primary importance. Recent studies have incorporated the similitude of the water film dynamics on the ice surface.[14]

[14] Alejandro Feo and Jen-Ching Tsao. The Water Film Weber Number in Glaze Icing Scaling. In *SAE Aircraft & Engine Icing International Conference*, number SAE-2007-01-3295, September 2007

ICING WIND TUNNEL INTER-COMPARISONS have been a convenient approach to further substantiate the scaling laws for ice accretion. Despite testing at presumably identical boundary conditions, huge discrepancies in ice shapes have been found in the first inter-comparison attempts; see Figure 2.5. Large measurement uncertainties in the LWC that explained the deviations were subsequently identified. The uncertainty of LWC measurement currently remains approximately 20%.

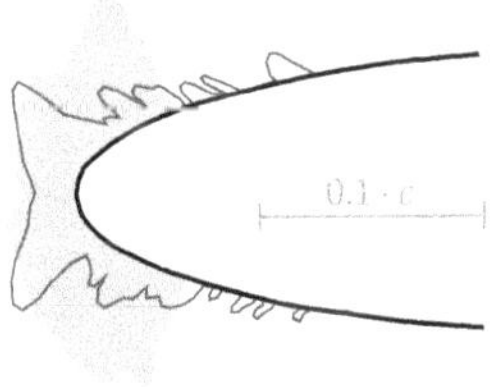

Figure 2.5: Icing wind tunnel inter-comparison. NACA 0012 aerofoil, AoA=0°, $c = 0.53\,\mathrm{m}$, $T_\infty = 264.8\,\mathrm{K}$, $U_\infty = 67\,\mathrm{m\,s^{-1}}$, MVD=40 µm, LWC≈ $0.61\,\mathrm{g\,m^{-3}}$, $t_{acc} = 11.2\,\mathrm{min}$. Image reproduced from BOND and ANDERSON (2004).

3
Selected physical properties of ice

Ice refers to the solid state of water. A short introduction to this chemical compound will elaborate its unusual properties. Water is the predominant naturally occurring inorganic liquid on Earth.[1] Its metastable state – supercooled water – is of paramount importance for aircraft icing. During the freezing process, supercooled water entraps air and the resulting ice exhibits a porous microstructure, which will be presented at the end of this chapter.

[1] Felix Franks. Water, 2000

3.1 Water and its different phases

The water molecule

A water molecule consists of two hydrogen atoms bound to one oxygen atom. Ten electrons are involved that pair up into five orbitals.[2] One electron pair remains attached to the oxygen nucleus. Two electron pairs create the two covalent bonds between oxygen and hydrogen. Another two electron pairs are grouped in an orbital around the oxygen nucleus on the far side of the covalent bond and are referred to as lone-pair orbitals. To satisfy a force equilibrium between these orbitals, the electron density distribution forms a rotund system in which the bonding angle between oxygen and hydrogen becomes 104.52°, see Figure 3.1.

[2] John A. Pople. The molecular orbital theory of chemical valency. V. The structure of water and similar molecules. In *Proceedings of the Royal Society of London A: Mathematical, Physical and Engineering Sciences*, volume 202, pages 323–336. The Royal Society, 1950

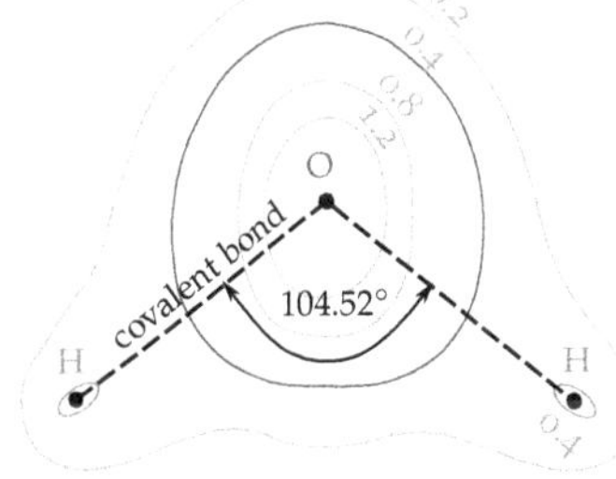

Figure 3.1: Approximated electron density contours of the water molecule. At equilibrium, the covalent bonds between oxygen and hydrogen form an angle of 104.52°. See Bader and Jones (1963) for further reading.

Hydrogen bonding and ice Ih

The majority of electrons are orbiting around the oxygen nucleus. Therefore, this region of the water molecule will be partially negatively charged. By contrast, the hydrogen nuclei are partially positively charged. Neighbouring water molecules will react based on this electric charge distribution. Indeed, the electrons in the lone-pair orbitals of one water molecule will attract the hydrogen nuclei of neighbouring water molecules and form a *hydrogen bond*.[3] At approximately 20 kJ mol^{-1}, the strength of a hydrogen bond is only about a twentieth of the strength of the covalent bond inside a water

[3] Wendell M. Latimer and Worth H. Rodebush. Polarity and ionization from the standpoint of the Lewis theory of valence. *Journal of the American Chemical Society*, 42(7):1419–1433, 1920

molecule, creating ideal conditions for promoting clustering between neighbouring water molecules.

THE ICE IH that we know from daily life as freezing water at atmospheric conditions demonstrates molecular clustering by hydrogen bonds; see Figure 3.2. The tetrahedral arrangement of one water molecule fosters four hydrogen bonds involving its two hydrogen atoms plus two further hydrogen bonds linking its lone-pair orbitals to the hydrogen atoms attached to adjacent water molecules. The entire three-dimensional cluster then forms hexagonal patterns, which is indicated by the letter 'h' in ice Ih.

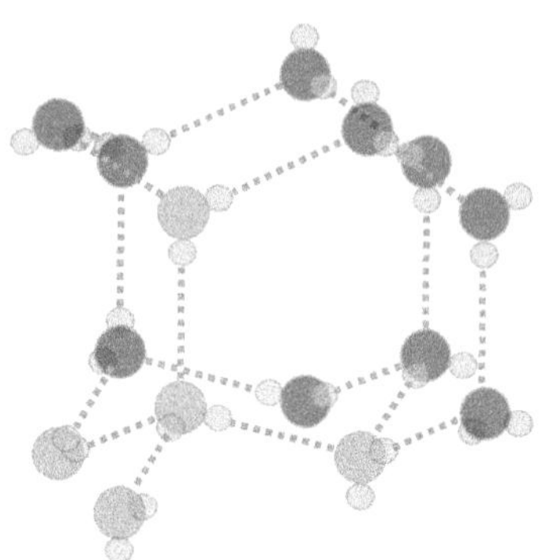

Figure 3.2: Hydrogen bonds (dotted lines) create regular patterns within ice Ih. Picture inspired by FLETCHER (1970).

Phase diagram of water

The Roman number I in ice Ih was assigned by TAMMANN, who pressurized ice up to 400 MPa and discovered structural changes in the material, which led to the discovery of ice II and ice III.[4] Over time, other phases of ice have been identified.[5] Figure 3.3 summarizes a small detail of the phase diagram of water. Clearly, atmospheric icing will involve ice Ih only. Therefore, the suffix Ih in later parts of this book will be omitted.

[4] Gustav Tammann. Über die Grenzen des festen Zustandes IV. *Annalen der Physik*, 307(5):1–31, 1900

[5] Victor F. Petrenko and Robert W. Whitworth. *Physics of ice*. Oxford University Press, 2006. ISBN 0198518943

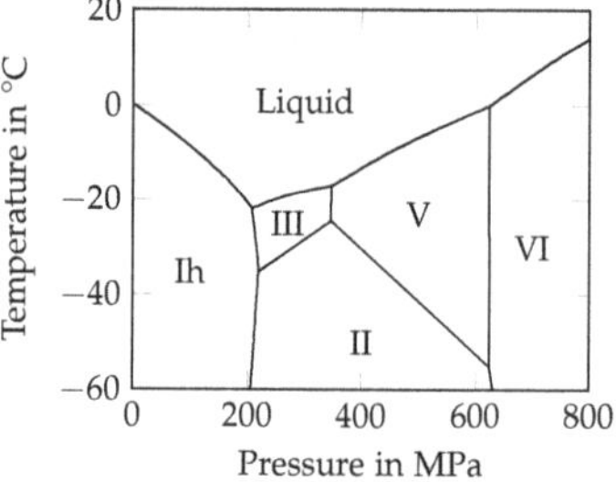

Figure 3.3: Phase diagram of water at equilibrium.

Supercooled water

The phase diagram of water suggests that under atmospheric conditions, liquid water will transform into ice as soon as the temperature drops below 0 °C. Anyone who has placed a bottle of water in the freezer and was surprised by the physical events thereafter knows differently; water can remain liquid even below 0 °C. This metastable state of water is called *supercooling*.[6] Astonishingly, a coherent theory of this phenomenon fully encompassing its thermodynamic and transport properties does not yet exist. A short outline of this difficult topic shall be given here.

THE HYDROGEN BOND is one of the underlying secrets.[7] Comparing its strength of 20 $kJ\,mol^{-1}$ with the latent heat of fusion of 6.01 $kJ\,mol^{-1}$ implies that only a minor fraction of the hydrogen bonds is affected by the solid/liquid phase change. Consequently, local tetrahedral symmetry of ice Ih, supercooled water and water close to its melting point occurs: in each of these cases, water molecules have joined to form clusters. The difference between these states lies in the spatial extension of the molecular order: liquid water has only a short-range order, while ice features a long-range crystalline order. To initiate a phase change, the transient tetrahedral clusters must group together and create a larger structure, an embryo. This process

[6] Charles A. Angell. Supercooled water. *Annual Review of Physical Chemistry*, 34 (1):593–630, 1983

[7] Pablo G. Debenedetti. Supercooled and glassy water. *Journal of Physics: Condensed Matter*, 15(45):R1669, 2003

is called nucleation and will be discussed in Chapter 4.1. Below the homogeneous nucleation temperature T_{homo} (approximately −40 °C at ambient pressure), the crystallization process becomes thermodynamically favourable, and ice Ih forms rapidly.

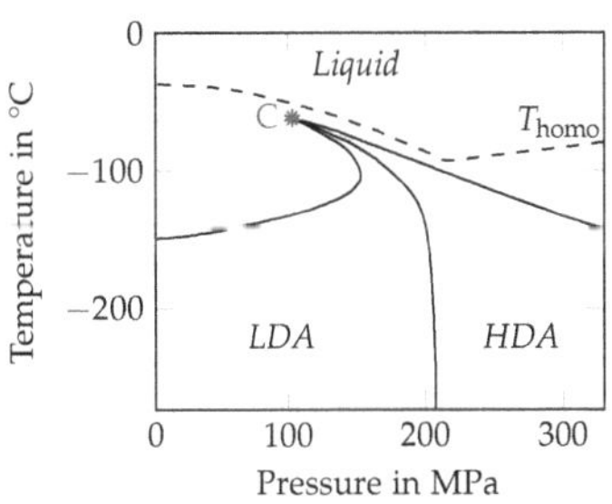

Figure 3.4: Phase relations of metastable water. Adapted by permission from Springer Nature from MISHIMA and STANLEY (1998).

WHEN LIQUID WATER IS RAPIDLY COOLED down below the glass transition temperature T_G (−140 °C at ambient pressure), crystallization is bypassed, and a metastable amorphous ice (also called glassy water) can exist. Although it is a solid material, its molecular structure suggests an arrangement of short-range order similar to that of a liquid. Depending on the ambient pressure, two amorphous forms are observed: low-density amorphous ice (LDA) and high-density amorphous ice (HDA).

MOLECULAR DYNAMICS STUDIES give reason to hypothesize that liquid water, supercooled water, LDA and HDA are related to each other;[8] see Figure 3.4. Between these metastable phases, a liquid-liquid phase transition is assumed, yielding a critical point C. At this critical point, the molecular states are fluctuating and might describe the anomalous behaviour of water. Currently, this region in the phase diagram is not accessible with experiments because the homogeneous nucleation process below T_{homo} is faster than current limits of experimental observation. Therefore, water is still a challenging field of study in condensed matter physics.

[8] Osamu Mishima and H. Eugene Stanley. The relationship between liquid, supercooled and glassy water. *Nature*, 396(6709):329–335, 1998

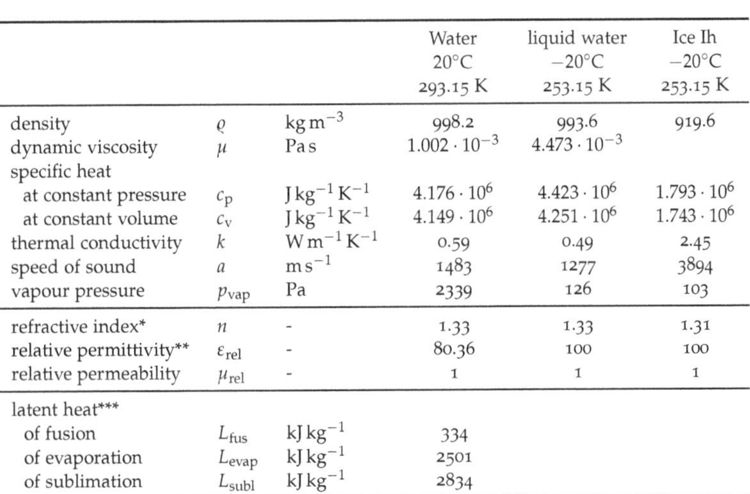

			Water 20°C 293.15 K	liquid water −20°C 253.15 K	Ice Ih −20°C 253.15 K
density	ϱ	$kg\,m^{-3}$	998.2	993.6	919.6
dynamic viscosity	μ	Pa s	$1.002 \cdot 10^{-3}$	$4.473 \cdot 10^{-3}$	
specific heat					
at constant pressure	c_p	$J\,kg^{-1}\,K^{-1}$	$4.176 \cdot 10^{6}$	$4.423 \cdot 10^{6}$	$1.793 \cdot 10^{6}$
at constant volume	c_v	$J\,kg^{-1}\,K^{-1}$	$4.149 \cdot 10^{6}$	$4.251 \cdot 10^{6}$	$1.743 \cdot 10^{6}$
thermal conductivity	k	$W\,m^{-1}\,K^{-1}$	0.59	0.49	2.45
speed of sound	a	$m\,s^{-1}$	1483	1277	3894
vapour pressure	p_{vap}	Pa	2339	126	103
refractive index*	n	-	1.33	1.33	1.31
relative permittivity**	ε_{rel}	-	80.36	100	100
relative permeability	μ_{rel}	-	1	1	1
latent heat***					
of fusion	L_{fus}	$kJ\,kg^{-1}$	334		
of evaporation	L_{evap}	$kJ\,kg^{-1}$	2501		
of sublimation	L_{subl}	$kJ\,kg^{-1}$	2834		

Table 3.1: Selected physical data on water, supercooled water and ice Ih. Data sources: BIDDLE et al. (2013), CHO et al. (1999), DASS and GILRA (1966), FEISTEL and WAGNER (2006), KEDENBURG et al. (2012), HAYNES (2016), HODGE and ANGELL (1978), MELINDER (2010), PERRY and GREEN (2008), VOGT et al. (2008), glossary.ametsoc.org, National Institute of Standards and Technology – Chemical WebBook at http://webbook.nist.gov. Annotations: *for visible light, **for low-frequency electromagnetic radiation, ***for a temperature of 0 °C.

3.2 *Data*

The anomalous behaviour of water is reflected in many of its properties; for instance, the density decreases when water freezes at 0 °C, whereas other liquids contract on freezing. Some other astonishing properties of water are summarized in Table 3.1, which conflates many references on quantitative data of liquid water, supercooled water and ice.

3.3 *Porosity of accreted ice*

Typical examples of transport phenomena in porous media are encountered in groundwater circulation, oil production, filtration processes, biological tissues and sound absorption. Depending on their composition, porous media will alter the surrounding fluid flow.[9] Does accreted ice on aircraft structures imply a porosity that affects the airflow?

[9] Jacob Bear. Introduction to Modeling of Transport Phenomena in Porous Media, 1990

To ANSWER THIS QUESTION, glaze ice formation[10] on a two-dimensional model of an engine nacelle inlet was analysed[11] by X-ray microtomography.[12] The experimental setup is illustrated in Figure 3.5. A small ice sample (8) is placed into a container (2), which is cooled by Peltier elements and a heat exchanger (3) to avoid degradation of the ice during the scan phase of 5 hours. The container is made out of boron nitride (7), which exhibits very low X-ray absorption. X-rays are emitted from a source (1,5) and pass through the ice sample, creating a characteristic radiograph. The absorption, density and shape of the ice sample will influence the radiograph. A scintillator (9) converts the ionizing radiation into visible light, which is projected on a CCD sensor (11) using a system of lenses (10). The sample container is placed on a rotating stage (4), thus enabling scans from multiple directions (12). The images are then reconstructed to obtain the three-dimensional microstructure of the glaze ice sample (13).

[10] The boundary conditions of the test case were as follows: $U_\infty = 40\,\mathrm{m\,s^{-1}}$, $T_\infty = 268\,\mathrm{K}$, LWC=$0.61\,\mathrm{g\,m^{-3}}$, MVD=$40\,\mu\mathrm{m}$, $t_{acc} = 300\,\mathrm{s}$.

[11] Juan S. Velandia, Alexander Diener, and Stephan E. Bansmer. Porosity of glaze ice from micro computed tomography. To be submitted to Journal of Glaciology

[12] Eric N. Landis and Denis T. Keane. X-ray microtomography. *Materials characterization*, 61(12):1305–1316, 2010

Figure 3.5: Experimental setup and data post-processing to measure the microscopic structure of glaze ice samples.

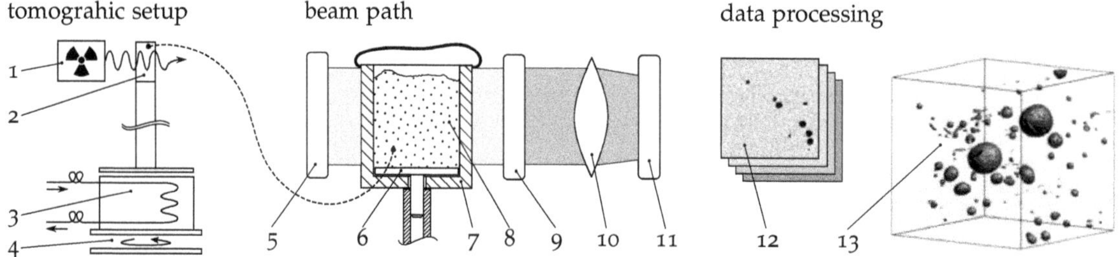

FURTHER POST-PROCESSING of the volumetric dataset reveals quantitative information on the porosity. Figure 3.6 presents a histogram of the equivalent pore diameter $d_{eq,\,pore}$. Because pores might exhibit an irregular shape, the $d_{eq,\,pore}$ is commensurate with a sphere of identical volume. The chart identifies many pores that are close to the resolution limit of the measurements. However, the sample's porosity of 1.2 %, defined as the ratio between the volume of confined air and the total volume of interest, is dominated by the few pores of size larger than 100 µm.

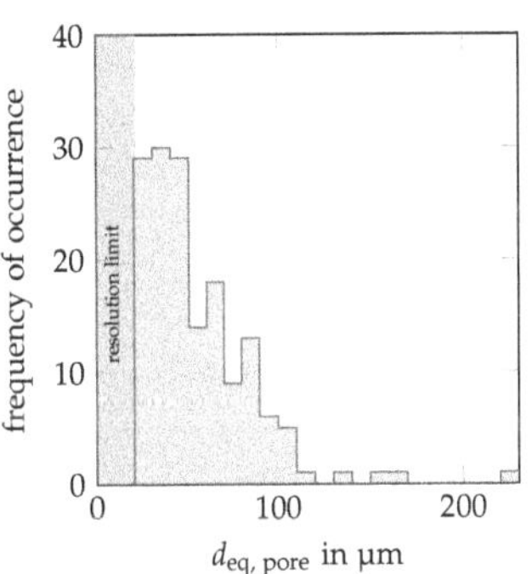

Figure 3.6: Size distribution of the air pores inside the glaze ice sample.

THE SHAPE OF THE PORES can be characterized by the sphericity $\odot$, which is given by

$$\odot = \frac{\pi^{\frac{1}{3}} \cdot (6V_{pore})^{\frac{2}{3}}}{A_{pore}} \quad ,$$

where V_{pore} and A_{pore} represent the pore volume and surface, respectively. A sphere has $\odot = 1$; the smaller the value of $\odot$, the less spherical the pores. Consequently, Figure 3.7 suggests that most of the entrapped air pores are spherical. Most likely, the freezing process of glaze ice accretion takes longer than the surface energy transfer to form air bubbles inside the liquid layer of supercooled water.

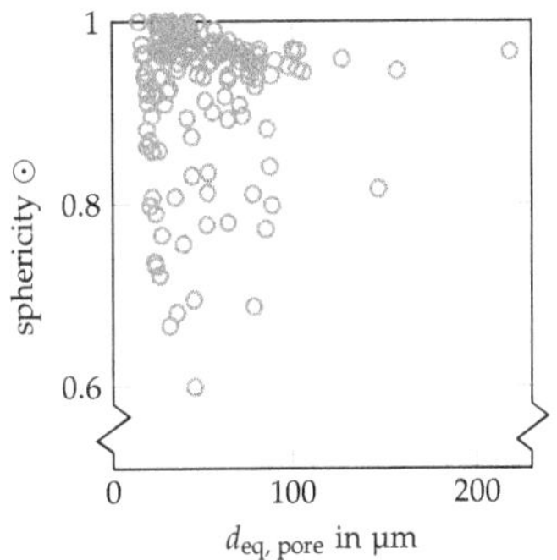

Figure 3.7: Scatter plot of the pore sphericity of the glaze ice sample.

Part II

Fluid mechanics of icing at small scales

Overview

The impact of supercooled droplets is the underlying primary mechanism of classic aircraft ice accretion at small scales. Two time scales must be considered. On the one hand, there is the kinetic time scale of droplet deformation, surface wetting and breakup into secondary droplets. On the other hand, the thermodynamic time scale of solidification competes with the kinetic phenomena. This part of the manuscript aims to introduce the complex physics of solidification and presents further studies on the droplet impact. Depending on whether the involved impact surface is dry or previously wetted, the outcome of the impingement process differs significantly and is hence strongly influential on the entire course of icing.

4
From nucleation to the solidification of impacting super-cooled droplets

Broad ranges of length and time scales are involved in the solidification process of impacting, supercooled droplets. Phase transition is initiated by small embryos of several nanometres in size and spreads through an entire droplet, which is in the millimetre size range in the case of freezing rain conditions. This size difference spans six orders of magnitude. In addition, the time scales of kinematic and thermodynamic phenomena vary from 10^{-6} to 1 seconds. Therefore, this chapter aims to introduce the major solidification mechanisms and motivate further drop impact studies for icing research.

4.1 *Nucleation*

Promoted by hydrogen bonding, both supercooled water and ice show certain similarities in their molecular clustering; see again the discussion in Chapter 3.1. The differences are due to the spatial extension of order: water has only short-range order, while ice features long-range crystalline order. In the very early period of the phase transition from the liquid to the solid state, self-organization occurs that locally extends the short-range order: a small nucleus of ice – the embryo – is formed. Subsequently, an interface separates the embryo from the surrounding liquid water. The kinetic mechanisms that ultimately lead to a phase transition of the entire water volume are collectively called *nucleation*.

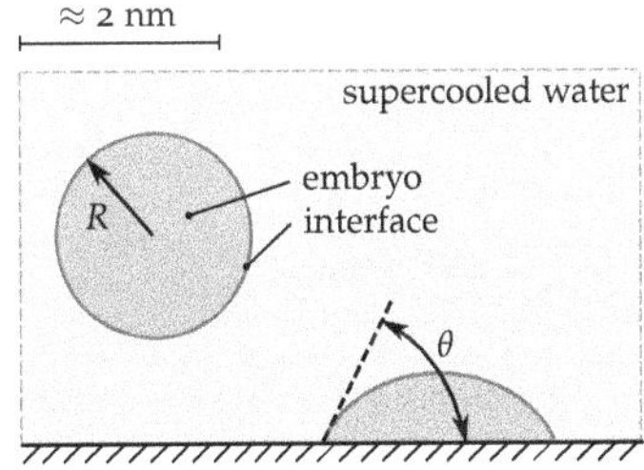

Figure 4.1: Model of an ice embryo in the bulk of supercooled water and on a flat surface according to classical nucleation theory.

Homogeneous nucleation

The emergence and growth of embryos within a bulk metastable phase of water in the absence of any suspended impurities is called *homogeneous* nucleation. Spontaneous microscopic fluctuations in the temperature, pressure or density trigger the formation of embryos.

Classical nucleation theory is a simplified approach to study the growth process of these embryos. The theory, which was originally developed to describe the condensation of a vapour to a liquid, can also be applied with some limitations to ice nucleation.[1] The basic idea is to consider the change in the free energy ΔG due to two competing mechanisms. On the one hand, energy is needed to increase the embryo surface A. On the other hand, the water molecules that have attached to the embryo and transitioned to the solid phase release energy while the embryo is growing in volume V. The total energy yields an equation for ΔG:

[1] Pablo G. Debenedetti. *Metastable liquids*. Physical chemistry. Princeton Univ. Press, 1996. ISBN 0691085951

$$\Delta G = \sigma \cdot \Delta A - G_v \cdot \Delta V \quad , \tag{4.1}$$

where σ is the surface tension and G_v characterizes the driving force of the phase change resulting from the lower free energy of the growing solid phase. Given the equilibrium melting temperature of ice $T_\mathrm{m} = 273.15\,\mathrm{K}$, the temperature T of the surrounding supercooled liquid, the latent heat of fusion L_fus, and the molecule number density ϱ_n of the solid phase, G_v becomes

$$G_v = \frac{\varrho_n \left(T_\mathrm{m} - T\right) L_\mathrm{fus}}{T_\mathrm{m}} \quad .$$

Assuming that the embryo is spherical in shape with radius R, one obtains

$$\begin{aligned}\Delta G &= \sigma \cdot 4\pi \left(R^2 - R_0^2\right) - G_v \cdot \frac{4}{3}\pi \left(R^3 - R_0^3\right) \\ &= \sigma \cdot 4\pi R^2 - G_v \cdot \frac{4}{3}\pi R^3 + \mathrm{const}\left(R_0\right) \quad ,\end{aligned}$$

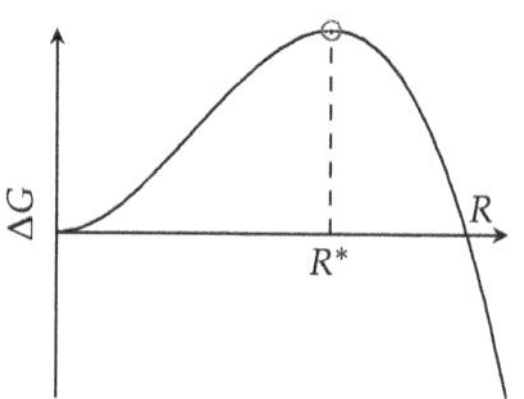

Figure 4.2: Change of the free energy ΔG of supercooled water due to the growth of an ice embryo.

where $R_0 \rightarrow 0$ indicates the initial radius of the embryo after its formation; Figure 4.2 plots this relation. The free energy of a small embryo first increases with the increasing size, thus impeding the growth process. Having reached a critical radius R^*, the embryo arrives at the nucleation barrier ΔG^*_homo, which is given by

$$\Delta G^*_\mathrm{homo} = \Delta G\left(R^*\right) = \frac{16}{3}\pi\frac{\sigma^3}{G_v^2} = \frac{16}{3}\pi\frac{\sigma^3 T_\mathrm{m}^2}{\varrho_n^2 \left(T_\mathrm{m} - T\right)^2 L_\mathrm{fus}^2} \quad .$$

This barrier has to be exceeded by the embryo to continue the phase transition process. Increasing the supercooling $(T_\mathrm{m} - T)$ lowers ΔG^*_homo, thus favouring phase transition.

Heterogeneous nucleation

When an embryo is formed and comes in contact with a surface, *heterogeneous* nucleation is triggered.[2] Typically, interaction between the surface and embryo significantly reduces the nucleation barrier of free energy. The simplest case of a surface, a smooth flat plate, is

[2] Richard P. Sear. Nucleation: theory and applications to protein solutions and colloidal suspensions. *Journal of Physics: Condensed Matter*, 19(3):033101, 2007

considered; see again Figure 4.1. When classical nucleation theory is applied, the nucleation barrier can be expressed as

$$\Delta G_{\text{hetero}} = \Delta G_{\text{homo}} \cdot f(\theta) \quad ,$$

where θ is the contact angle between the surface and the embryo interface. A dominating attraction between the molecules of the embryo and the surface yield a thin-spread embryo with a small contact angle. Otherwise, the embryo is repulsed from the surface, and θ approaches 180°. The function $f(\theta)$ is plotted in Figure 4.3. As $\theta \to 0$, $f \to 0$, implying a vanishing nucleation barrier. For large contact angles, f approaches the value of 1, which corresponds to the case of homogeneous nucleation, because $\Delta G_{\text{hetero}} \to \Delta G_{\text{homo}}$. According to the above equation, the presence of a surface thus decreases the nucleation barrier because the intermolecular attraction between the embryo and surface makes it easier for the embryo to spread and create the new surfaces that are needed for nucleation. This process also promotes a phase transition with a shorter time delay relative to the case of homogeneous nucleation. Rough surfaces further increase this tendency.

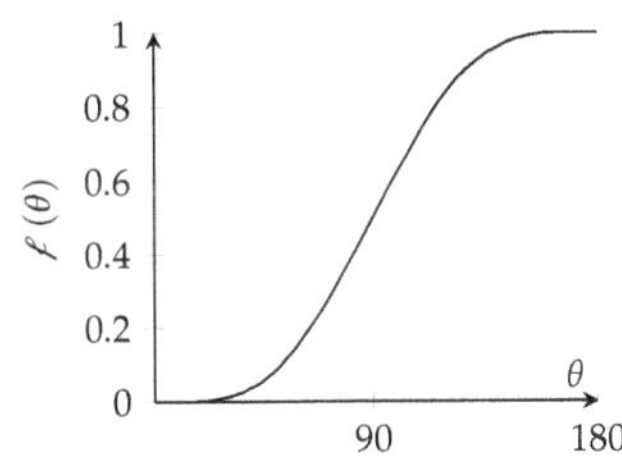

Figure 4.3: Function $f(\theta)$ that relates the homogeneous and heterogeneous nucleation barriers.

ONE OF THE APPROXIMATIONS made by classical nucleation theory is to assume that the small embryos have the bulk phase properties of ice. Thereby, statistical mechanics at a molecular level are circumvented; however, only the general trends of nucleation can be explained through this approach. Density functional theory[3] and molecular simulations[4] are modern tools to study nucleation in more detail.

[3] David W. Oxtoby. Density functional methods in the statistical mechanics of materials. *Annual Review of Materials Research*, 32(1):39–52, 2002

[4] Peng Yi and Gregory C. Rutledge. Molecular origins of homogeneous crystal nucleation. *Annual Review of Chemical and Biomolecular Engineering*, 3: 157–182, 2012

4.2 *Dendrites*

The growing ice embryo, still spherical in shape, binds neighbouring clusters of water molecules because the formation of long-range crystalline order is favoured. Such binding causes a release of latent heat in the vicinity of the embryos' surface, which has to be dissipated into the surrounding supercooled water, mainly by thermal diffusion.[5] To maximize the heat transfer, the embryo tends to increase its surface area as much as possible. Therefore, a MULLINS-SEKERKA instability[6] is triggered that destabilizes the spherical shape of the embryo. Harmonic shape variations – small ripples – emerge. Based on a linear stability analysis of the ice-water interface and assuming the thermal diffusion of the latent heat, one can show that this change in the interfacial morphology is initiated when the radius of the embryo is approximately seven times the critical radius R^* of the nucleation barrier. Depending on the degree of supercooling, this

[5] James S. Langer. Instabilities and pattern formation in crystal growth. *Reviews of Modern Physics*, 52(1):1, 1980

[6] William W. Mullins and Robert F. Sekerka. Morphological stability of a particle growing by diffusion or heat flow. *Journal of Applied Physics*, 34(2): 323–329, 1963

critical value corresponds to an embryo size of approximately 30 nm. Subsequently, the surface ripples evolve into a needle crystal. Ignoring the effect of surface tension, theory[7] predicts that the needles will develop a parabolic shape. Further maximizing the interfacial area between the ice and water for efficient heat transfer, the parabola creates side branches, corresponding to dendrite formation; see Figure 4.4. The Greek etymology of dendrites ($\delta\epsilon\nu\delta\varrho\acute{\iota}\tau\eta\varsigma$) accurately describes their shape: tree-like.

[7] Georgy P. Ivantsov. Temperature around a spheroidal, cylindrical and acicular crystal growing in a supercooled melt. In *Dokl. Akad. Nauk.*, volume 58, pages 567–569, 1947

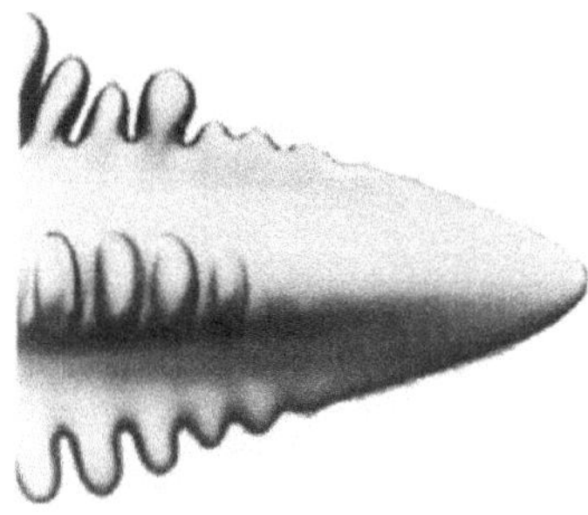

Figure 4.4: Dendritic tip structures in succinonitrile ($C_4H_4N_2$). Original picture from GLICKSMAN (1984) with further image processing. Reprinted with permission from Elsevier.

4.3 *Freezing of a sessile droplet*

Having discussed microscopic concepts of solidification, we consider the events on a macroscopic scale for the case of a sessile droplet. As an approximation, the sessile droplet has the shape of a spherical cap, which impedes the visualization of its solidification significantly: incoming parallel light rays are scattered by this geometry in various directions; see also the explanations on page 77.

Hele-Shaw cell for experimental investigations

The Hele-Shaw cell was originally designed to visualize the streamlines of two-dimensional flow.[8] This device consists of two parallel, transparent plates separated by a slight distance. The fluid motion between these plates is dominated by viscous effects and shows similarities to two-dimensional potential flow.[9] Due to capillary forces, a sessile droplet placed in a Hele-Shaw cell spontaneously spreads into a nearly two-dimensional circular segment. The light scattering of the incident rays perpendicular to the two plates is thus minimized.

[8] Henry Selby Hele-Shaw. Investigation of the nature of surface resistance of water and of stream-line motion under certain experimental conditions. volume XL, page 25. Trans. Inst. Nav. Arch., 1898

[9] George Gabriel Stokes. Mathematical proof of the identity of the stream lines obtained by means of a viscous film with those of a perfect fluid moving in two dimensions. *Brit. Ass. Rep*, 143, 1898

Freezing phenomenology

The freezing of a sessile droplet in a Hele-Shaw cell is illustrated in Figure 4.5.[10] A copper plate, mounted at the bottom, is used to initially supercool the droplet to a temperature T_∞. Two distinct solidification phases can be observed. During the *first phase*, heterogeneous nucleation is triggered at the copper surface. An array of dendrites

[10] Markus Schremb and Cameron Tropea. Solidification of supercooled water in the vicinity of a solid wall. *Physical Review E*, 94:052804, Nov 2016

Figure 4.5: Solidification of a supercooled drop within a Hele-Shaw cell. Adapted from SCHREMB and TROPEA (2016) with additional image post-processing with permission from the American Physical Society.

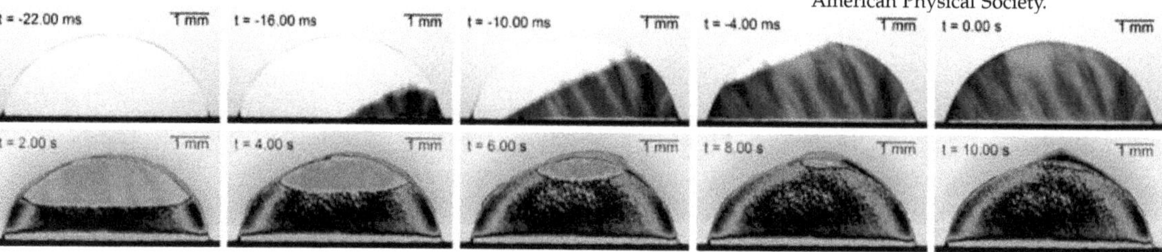

starts to grow that is clearly visible in the upper row of Figure 4.5. Their tip velocity can be estimated to be $0.17\,\mathrm{m\,s^{-1}}$. The multitude of dendrites promotes directional growth because direct neighbours restrict dendritic side-branching. After 20 ms, the dendrites have traversed the entire droplet volume. Simultaneously, the latent heat is released into the remaining supercooled water, which prevents the droplet from becoming fully frozen. These effects occur during the *second solidification phase*; see the lower row of Figure 4.5. The copper surface remains at a low temperature and thus acts as a heat sink. In the opposite direction of the heat flux, the solidification front traverses the droplet with a relatively low velocity of approximately $0.25 \cdot 10^{-3}$ m/s. Since the density of ice is lower than that of water, the droplet volume is increasing, ultimately yielding a cusp-like shape at its upper end.

Temperature evolution

The two solidification phases also cause changes in the droplet temperature; see Figure 4.6.[11] Starting with T_∞ at the supercooled state, the rapid dendrite growth releases latent heat, which increases the temperature. When the melting temperature T_m is reached, the supercooling capacity is completely exhausted, and the dendritic growth is terminated. At this stage, the droplet consists of both water and ice, with its exact volume ratio depending on the supercooling. While the heat is conducted from the droplet into the copper substrate during the second solidification phase, the droplet temperature remains at T_m. After the droplet is fully frozen, its temperature decreases again towards T_∞.

[11] Jason P. Hindmarsh, Alexander B. Russell, and Xiao D. Chen. Experimental and numerical analysis of the temperature transition of a suspended freezing water droplet. *International Journal of Heat and Mass Transfer*, 46(7): 1199–1213, 2003

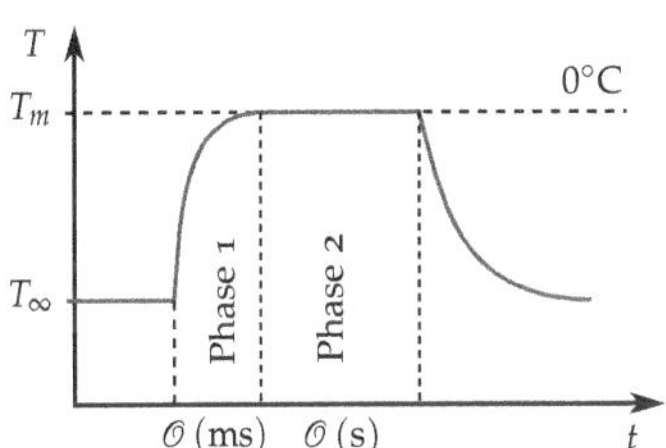

Figure 4.6: Temperature evolution of a freezing droplet that was initially supercooled.

4.4 *Impact of supercooled droplets*

In the case of aircraft icing, the supercooled droplets are not sessile but are instead impinging on the aerodynamic surfaces, which raises an important question: do the droplets fully freeze during impact?

Two different timescales are compared in the following: a *kinetic timescale* t_kin that characterizes the impact event and a *solidification timescale* t_solid. Assume a spherical droplet with diameter D and an impact velocity w_imp. A kinetic reference time can be defined as $t_\mathrm{kin,ref} = \frac{D}{w_\mathrm{imp}}$. The temporal evolution of the droplet impact can then be considered as a multiple $\mathfrak{t}$ of the reference time

$$t_\mathrm{kin} = \mathfrak{t} \cdot t_\mathrm{kin,ref} = \mathfrak{t} \cdot \frac{D}{w_\mathrm{imp}} \quad .$$

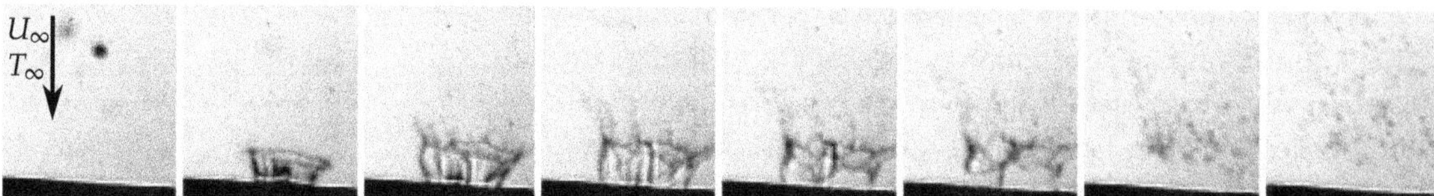

Figure 4.7: Icing observed in an icing wind tunnel with a high magnification: an impacting supercooled droplet with a diameter of approximately 60 µm. $U_\infty = 40\,\mathrm{m\,s^{-1}}$, $T_\infty = -5\,°\mathrm{C}$. Data provided by ARNE BAUMERT.

After contacting the aerodynamic surface, the droplet spreads and forms a thin liquid film with film height h_{film}. Subsequently, splashing[12] might occur; see page 51 for more detailed descriptions of the physical phenomena. The solidification timescale of this thin film is influenced by the growth velocity of the dendrites w_{den}, yielding

$$t_{\text{solid}} = \frac{h_{\text{film}}}{w_{\text{den}}} = \frac{\mathfrak{h}_{\text{film}} \cdot D}{w_{\text{den}}} \quad ,$$

where $\mathfrak{h}_{\text{film}} = \frac{h_{\text{film}}}{D}$ represents the non-dimensional film thickness. Supposing that the droplet remains liquid during impact, the kinetic process must be more rapid than the solidification; hence,

$$\begin{aligned} t_{\text{kin}} &\leq t_{\text{solid}} \\ \mathfrak{t} \cdot \frac{D}{w_{\text{imp}}} &\leq \frac{\mathfrak{h}_{\text{film}} \cdot D}{w_{\text{den}}} \\ \mathfrak{t} &\leq \mathfrak{h}_{\text{film}} \cdot \frac{w_{\text{imp}}}{w_{\text{den}}} \quad . \end{aligned}$$

Estimating the boundary conditions of aircraft icing, the impact kinetics proceed until the non-dimensional time becomes

$$\mathfrak{t} = 0.01 \cdot \frac{100\,\mathrm{m\,s^{-1}}}{0.17\,\mathrm{m\,s^{-1}}} \approx 6 \quad .$$

At this stage, the impact process has significantly advanced – an undisturbed droplet would have travelled 6 times its diameter. In case of splashing, secondary droplets have already been created.

THE HIGH-SPEED VISUALIZATION of the process of aircraft icing at small scales was performed recently; see Figure 4.7. The image sequence is obtained using long-distance microscopy and shows a supercooled droplet that impacts on a hemisphere. Splashing can be clearly observed, i.e., the droplet has not solidified during its impact. In the following chapters, therefore, I seek to study the droplet impact at room temperature, neglecting any phase change.

[12] Splashing refers to a morphological change of the impacting primary droplet that ultimately results in the generation of small secondary droplets.

5
Methods to study drop impact at ambient conditions

Aircraft icing involves droplet impacts at high velocities. This scenario poses several challenges for both experimental and computational studies. An experimental realization under laboratory conditions cannot be achieved by a pure gravitational acceleration of the droplet – aerodynamic drag limits the droplet's fall velocity. Computational tools are used to model the liquid/gas interface that contains length scales spanning approximately three orders of magnitude. The presented methods form the basis of the subsequent investigations on droplet impact.

5.1 Boundary conditions

Interface and surface tension

An *interface* is a region in space that separates two immiscible fluids. The early theories of YOUNG, LAPLACE and GAUSS considered the interface to have zero thickness, yielding discontinuous changes in macroscopic properties such as density and viscosity when crossing it. Subsequently, it was recognized by MAXWELL, GIBBS and VAN DER WAALS that the interface has a finite thickness, where smooth yet rapid transitions of the macroscopic properties occur.[1]

[1] Daniel M. Anderson, Geoffrey B. McFadden, and Adam A. Wheeler. Diffuse-interface methods in fluid mechanics. *Annual review of fluid mechanics*, 30(1):139–165, 1998a

A LIQUID-GAS INTERFACE on an idealized molecular level, neglecting the violent kinetic agitation in that region, is shown in Figure 5.1. For molecules in the liquid bulk, intermolecular forces compensate for each other in all directions. When approaching the interface, this force balance is altered: a molecule of the liquid phase, which has the tendency to escape the bulk in that region, is drawn back. The interface thus exhibits a resistance against being stretched, which is characterized by the *surface tension* σ. To change the shape of the interface by increasing its area A_I, work has to be performed that relates to a change in free energy G, yielding another definition for the surface tension:

$$\sigma = \left(\frac{\partial G}{\partial A_\mathrm{I}}\right)_{T,p,\mu=\mathrm{const}} ,$$

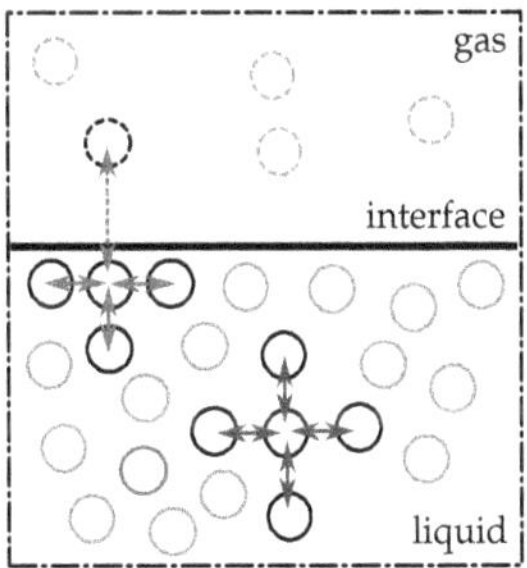

Figure 5.1: Simplified model of the molecular origin of surface tension.

which has already been used to describe the classical nucleation theory in equation (4.1) on page 38. The definition of surface tension has been refined over time;[2] modern theories are based on statistical mechanics concepts.[3]

[2] Yves Pomeau. *Surface tension: from fundamental principles to applications in liquids and in solids.* PhD thesis, Warsaw School of Statistical Physics, 2013

[3] John G. Kirkwood and Frank P. Buff. The statistical mechanical theory of surface tension. *The Journal of Chemical Physics*, 17(3):338–343, 1949

Dimensionless groups

The boundary conditions of a liquid spherical drop – surrounded by gas – impacting perpendicularly on a flat surface shall be defined. A drop with diameter D impacts with a velocity w_{imp}. Both the liquid and gas phases can be characterized by their densities ϱ_{liq} and ϱ_{gas} and dynamic viscosities μ_{liq} and μ_{gas}. The interface involves a surface tension σ. Inertial forces typically dominate over gravitational forces, which can thus be neglected. By applying the Buckingham Π theorem,[4] important dimensionless groups can be derived, namely, the Reynolds and Weber numbers

[4] Edgar Buckingham. On physically similar systems; illustrations of the use of dimensional equations. *Physical review*, 4(4):345, 1914

$$\mathrm{Re} = \frac{\varrho_{\mathrm{liq}} D w_{\mathrm{imp}}}{\mu_{\mathrm{liq}}} \qquad \mathrm{We} = \frac{\varrho_{\mathrm{liq}} D w_{\mathrm{imp}}^2}{\sigma} \quad ,$$

that quantify the balances between the inertial and viscous forces and the inertial and capillary forces, respectively. The viscosity ratio between the gas and liquid phases is conveniently introduced by the Stokes number

$$\mathrm{St} = \frac{\mu_{\mathrm{gas}}}{\mu_{\mathrm{liq}}} \frac{1}{\mathrm{Re}} \quad ,$$

which becomes significant in the initial phase of droplet impact when a thin gas layer is entrapped between the liquid drop and solid substrate. See also the later discussions in Chapter 6.

Drop impact conditions for aircraft icing

Aircraft operation during holding patterns and approach is the most challenging with respect to icing. The major cloud systems containing supercooled droplets are located at these flight levels, and the engines that are typically powering the de/anti-icing systems are set to low power. Table 5.1 exemplifies such scenarios and derives the boundary conditions for the droplet impact in terms of the Reynolds, Weber and Stokes numbers. Weber numbers beyond 20,000 should not considered for impact studies because the gas flow around the

Scenario	w_{imp} $\mathrm{m\,s^{-1}}$	D µm	T_∞ °C	Re	We	St
Heathrow holding, App. C	128.6	40	-14.7	1500	8500	$3.2 \cdot 10^{-6}$
Heathrow approach, App. O	82.3	200	-5.0	7600	18000	$1.0 \cdot 10^{-6}$

Table 5.1: Estimation of the dimensionless groups of droplet impact for typical icing scenarios under EASA certification specification CS 25, appendices C and O. Material properties from Hrubý *et al.* (2014) and Hallet (1963).

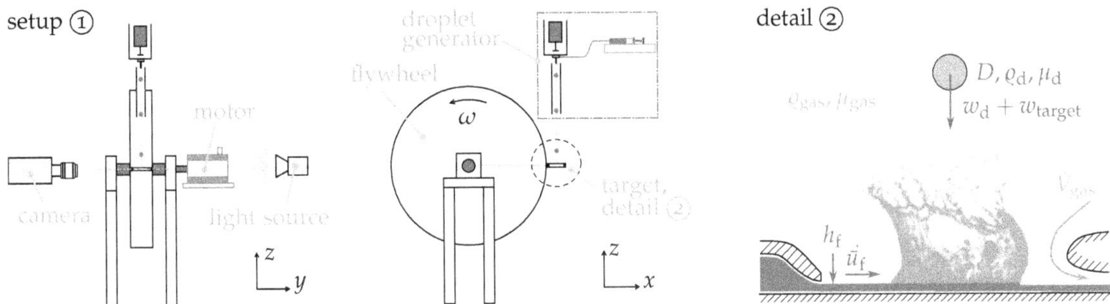

Figure 5.2: Experimental configuration: a target mounted on a flywheel provides high impact velocities. The target can be wetted to study drop impact on thin films.

droplet destabilizes its spherical shape and ultimately produces secondary breakup[5] before the droplet even hits the surface.

[5] Lien-Peng Hsiang and Gerard M. Faeth. Near-limit drop deformation and secondary breakup. *International Journal of Multiphase Flow*, 18(5):635–652, 1992

5.2 *A rotating rig for experimental drop impact studies*

The drop impact was studied experimentally. In this work, to emulate high impact velocities and thus high Weber numbers, a rotating rig is used;[6] see Figure 5.2. This rig consists of a target mounted on a flywheel that is rotating with an angular frequency ω. An on-demand droplet generator is synchronized with the flywheel motion and releases a drop that is accelerated by gravity. The flywheel rotates around the falling droplet several times until they collide with a total impact velocity of $w_{\text{imp}} = w_{\text{d}} + w_{\text{target}}$. Furthermore, the target can be wetted[7] with a thin film of height h_{f} and mean velocity $\bar{u}_{\text{f}}$. To maintain stable wetting, the film is sucked away at the outer end of the target. A shadowgraph configuration, composed of a light source and a camera, captures images during drop impact.

[6] Benjamin W. Faßmann, Stephan E. Bansmer, Thorsten J. Möller, Rolf Radespiel, and Michael Hartmann. High velocity impingement of single droplets on a dry smooth surface. *Experiments in fluids*, 54(5):1516, 2013

[7] David A. Burzynski and Stephan E. Bansmer. Droplet splashing on thin moving films at high Weber numbers. *International Journal of Multiphase Flow*, 101:202–211, 2018

5.3 *Direct numerical simulations of two-phase flow*

Computational studies complemented the experimental approach. Using a whole-domain formulation,[8] the incompressible, viscous, immiscible two-phase flow of an impacting droplet is governed by the continuity equation

$$\nabla \cdot \vec{u} = 0 \quad , \tag{5.1}$$

and the momentum balance

$$\frac{\partial (\varrho \vec{u})}{\partial t} + \nabla \cdot (\varrho \vec{u} \otimes \vec{u}) = \nabla p + \nabla \cdot \left[\mu \left(\nabla \vec{u} + (\nabla \vec{u})^{\mathrm{T}} \right) \right] + \varrho \vec{g} + \vec{f}_{\sigma} \ , \tag{5.2}$$

where $\vec{u} = (u, v, w)^{\mathrm{T}}$ is the velocity[9], p the pressure and $\vec{g}$ the gravitational acceleration. The surface tension force that acts at the liquid/gas interface is introduced as a continuous volumetric force $\vec{f}_{\sigma}$ according to the model of Brackbill.[10] Both the density ϱ and dy-

[8] Ruben Scardovelli and Stéphane Zaleski. Direct numerical simulation of free-surface and interfacial flow. *Annual Review of Fluid Mechanics*, 31(1):567–603, 1999

[9] $\vec{u} \otimes \vec{u}$ is the dyadic product of the velocity vectors yielding a matrix.

[10] Jeremiah U. Brackbill, Douglas B. Kothe, and Charles Zemach. A continuum method for modeling surface tension. *Journal of Computational Physics*, 100(2):335–354, 1992

namic viscosity μ are constant quantities but depend on the location of the interface as follows:

$$\varrho \simeq \begin{cases} \varrho_{\text{liq}} & \text{in the liquid} \\ \varrho_{\text{gas}} & \text{in the gas} \end{cases} \qquad \mu \simeq \begin{cases} \mu_{\text{liq}} & \text{in the liquid} \\ \mu_{\text{gas}} & \text{in the gas} \end{cases} \quad . \tag{5.3}$$

The direct numerical simulation of these equations faces two challenges that are discussed in the following sections. First, both the shape and kinematics of the liquid/gas interface are modelled to localize the density and viscosity and to determine the surface tension force $\vec{f}_\sigma$. Second, the so-called pressure-velocity coupling in equation (5.2) is addressed.

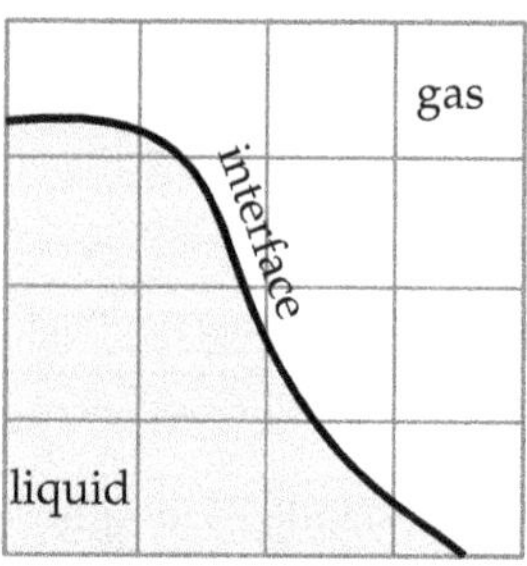

Figure 5.3: The VOF method uses a scalar indicator $0 \leq \chi \leq 1$ in each computational cell for implicit interface capturing.

The volume-of-fluid (VOF) method to model the interface

To capture the liquid/gas interface in the two-phase flow simulation, the VOF method[11] is used; see Figure 5.3. In each cell of the discretized computational domain, the liquid volume fraction is stored as a scalar indicator χ, thus yielding an implicit description of the interface:

$$\chi = \begin{cases} 1 & \text{if the cell is fully occupied by the liquid phase,} \\ 0 & \text{if the cell is fully occupied by the gas phase,} \\ 0 < \chi < 1 & \text{if the cell is in the vicinity of the interface.} \end{cases}$$

When the distribution of χ is known, the approximate relationships (5.3) for the density and viscosity can be refined as follows:

$$\varrho = \chi \varrho_{\text{liq}} + (1 - \chi)\, \varrho_{\text{gas}}$$
$$\mu = \chi \mu_{\text{liq}} + (1 - \chi)\, \mu_{\text{gas}} \quad .$$

Moreover, the surface tension force can be directly computed by

$$\vec{f}_\sigma = \sigma \nabla \cdot \left(\frac{\nabla \chi}{|\nabla \chi|} \right) \nabla \chi \quad .$$

Interface kinematics and the problem of numerical diffusion

The governing flow equations require a model delineating the temporal evolution of the interface, which is transported with the fluid velocity $\vec{u}$. Hence, the scalar indicator χ must satisfy the convection equation

$$\frac{\partial \chi}{\partial t} + \nabla \cdot (\chi \vec{u}) = 0 \quad .$$

Despite the apparent simplicity of this equation, its adequate discretization is ultimately very challenging. As a first example, consider a one-dimensional reduction of the above transport equation

[11] Cyril W. Hirt and Billy D. Nichols. Volume of fluid (VOF) method for the dynamics of free boundaries. *Journal of Computational Physics*, 39(1):201–225, 1981

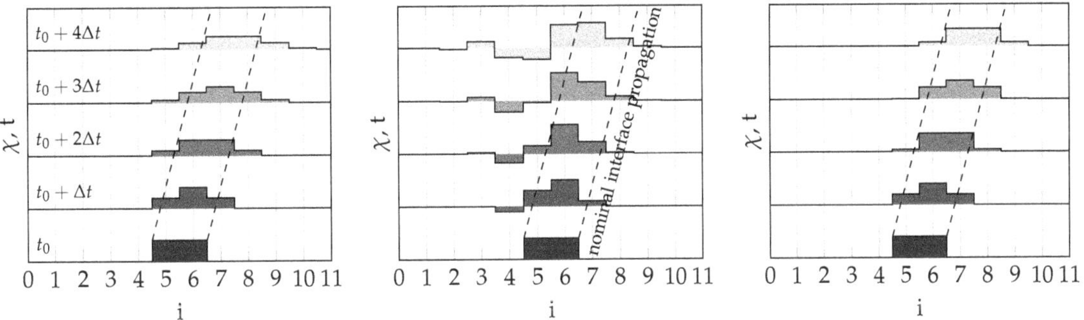

Figure 5.4: Comparison of different spatial discretization schemes for the VOF transport equation. *Left:* backward difference. *Centre:* central difference. *Right:* TVD scheme with VAN LEER limiter.

assuming a constant velocity u:

$$\frac{\partial \chi}{\partial t} + u \cdot \frac{\partial \chi}{\partial x} = 0 \tag{5.4}$$

with a first-order backward discretization in space and a first-order forward difference in time:

$$\frac{\chi_i^{n+1} - \chi_i^n}{\Delta t} + u \cdot \frac{\chi_i^n - \chi_{i-1}^n}{\Delta x} = 0 \quad .$$

The subscripts $i, i-1$ indicate discrete spatial coordinates separated by a distance Δx, and the superscripts $n, n+1$ represent time levels with an increment Δt. Substituting the Taylor expansions

$$\chi_{i-1}^n = \chi_i^n - \Delta x \left(\frac{\partial \chi}{\partial x}\right)_i^n + \frac{\Delta x^2}{2} \left(\frac{\partial^2 \chi}{\partial x^2}\right)_i^n - \cdots$$

$$\chi_i^{n+1} = \chi_i^n + \Delta t \left(\frac{\partial \chi}{\partial t}\right)_i^n + \frac{\Delta t^2}{2} \left(\frac{\partial^2 \chi}{\partial t^2}\right)_i^n + \cdots$$

into the discretized convection equation

$$\left(\frac{\partial \chi}{\partial t}\right)_i^n + u \cdot \left(\frac{\partial \chi}{\partial x}\right)_i^n = u \frac{\Delta x}{2} \left(\frac{\partial^2 \chi}{\partial x^2}\right)_i^n - \frac{\Delta t}{2} \left(\frac{\partial^2 \chi}{\partial t^2}\right)_i^n + \cdots,$$

and performing further algebra yields

$$\left(\frac{\partial \chi}{\partial t}\right)_i^n + u \cdot \left(\frac{\partial \chi}{\partial x}\right)_i^n = \frac{u \Delta x}{2} \left(1 - \frac{\Delta t}{\Delta x} u\right) \left(\frac{\partial^2 \chi}{\partial x^2}\right)_i^n + \cdots \quad .$$

This equation is the *equivalent differential equation* to (5.4). On the left-hand side, the convection property is represented, but the right-hand side does not vanish. Instead, due to a truncation error of the discretization scheme, a second-order derivative in space emerges that represents a diffusion mechanism. To visualize this phenomenon, an initial step function for χ at $t = t_0$ is incorporated using the above discretization; see Figure 5.4 (left). The velocity is set to $u = 0.5 \frac{\Delta x}{\Delta t}$.

Clearly, the sharp boundaries of the interface smear out when marching forward in time. The introduced numerical viscosity

$$\nu_{\text{numerical}} = \frac{u\Delta x}{2}\left(1 - \frac{\Delta t}{\Delta x}u\right)$$

is positive in sign for $0 \leq \frac{\Delta t}{\Delta x}u < 1$, thus damping the solution over time. If a central scheme of second-order accuracy for the spatial discretization is used, the solution becomes unconditionally unstable and creates new extrema; see Figure 5.4 (centre).

INDEED, GODUNOV'S THEOREM shows that any linear numerical scheme for the convection equation that is monotonic (i.e., does not create new extrema) can be at most first-order accurate.[12] All linear schemes of higher order cannot be monotonicity preserving. To circumvent this theorem, high-resolution schemes must involve nonlinearity. In that case, undesired oscillations of the solution can be suppressed by keeping the total variation bounded.[13] A total variational diminishing (TVD) scheme for the convection equation (5.4) can be composed as

$$\frac{\chi_i^{n+1} - \chi_i^n}{\Delta t} + u \cdot \left[1 + \frac{1}{2}\frac{\mathscr{L}\left(\mathfrak{r}_{i+\frac{1}{2}}\right)}{\mathfrak{r}_{i+\frac{1}{2}} + \epsilon} - \frac{1}{2}\mathscr{L}\left(\mathfrak{r}_{i-\frac{1}{2}}\right)\right] \cdot \frac{\chi_i^n - \chi_{i-1}^n}{\Delta x} = 0,$$

with

$$\mathfrak{r}_{i+\frac{1}{2}} = \frac{\chi_i^n - \chi_{i-1}^n}{\chi_{i+1}^n - \chi_i^n + \epsilon} \qquad \mathfrak{r}_{i-\frac{1}{2}} = \frac{\chi_{i-1}^n - \chi_{i-2}^n}{\chi_i^n - \chi_{i-1}^n + \epsilon} \qquad \epsilon = 10^{-5}$$

and a VAN LEER limiter[14] given by

$$\mathscr{L}(\mathfrak{r}) = \frac{\mathfrak{r} + |\mathfrak{r}|}{1 + |\mathfrak{r}|} \quad .$$

Figure 5.4 (right) confirms the satisfactory transport properties of the TVD discretization while minimizing the numerical diffusion.

[12] Sergei Konstantinovich Godunov. A difference method for numerical calculation of discontinuous solutions of the equations of hydrodynamics. *Matematicheskii Sbornik*, 89(3):271–306, 1959

[13] Ami Harten. High resolution schemes for hyperbolic conservation laws. *Journal of Computational Physics*, 49(3): 357–393, 1983

[14] Bram Van Leer. Towards the ultimate conservation difference scheme. II. Monotonicity and conservation combined in a second-order scheme. *Journal of Computational Physics*, 14:361–370, 1974

Pressure-velocity coupling

The pressure and density of a fluid are typically linked by an equation of state. For modelling incompressible flow, the usage of this relation is inappropriate because a variation of the pressure in space shall be allowed while restricting the density to be constant. In consequence, the incompressible continuity equation (5.1) loses both its time dependence and its strong coupling with the momentum equation (5.2) and deteriorates to a constraint on the velocity $\vec{u}$, which has to be satisfied through the correct choice of the pressure gradient ∇p in the momentum equation.[15] Indeed, the two unknowns – the pressure p and velocity $\vec{u}$ – have to be extracted mainly from the momentum equation.

[15] George D. Raithby and Gerald E. Schneider. Numerical solution of problems in incompressible fluid flow: treatment of the velocity-pressure coupling. *Numerical Heat Transfer, Part A: Applications*, 2(4):417–440, 1979

MANY ALGORITHMS, including SIMPLE[16] (Semi-Implicit Method for Pressure-Linked Equations) and PISO[17] (Pressure-Implicit with Splitting of Operators), have been introduced to address the above problem of pressure-velocity coupling. These algorithms solve segregated equations for the velocity and pressure. For each time step of the two-phase flow solver, the following simplified sequence is executed:

- Solve the interface equations, and calculate the density and viscosity fields.
- *Predictor Step.* Solve the momentum equation (5.2) using the pressure field from the previous iteration. An approximation of the new velocity field $\vec{u}^{\star}$ is obtained.
- *First corrector step.* Take the divergence of the momentum equation, and combine it with the continuity equation to obtain the pressure distribution $p^{\star\star}$, assuming $\vec{u}^{\star}$ for the velocity.
- *Second corrector step.* Use the new pressure distribution $p^{\star\star}$ to update the velocity field.
- Iterate until a pre-determined tolerance is reached.

[16] Suhas V. Patankar and D. Brian Spalding. A calculation procedure for heat, mass and momentum transfer in three-dimensional parabolic flows. *International Journal of Heat and Mass Transfer*, 15(10):1787–1806, 1972; and Suhas V. Patankar. A calculation procedure for two-dimensional elliptic situations. *Numerical Heat Transfer*, 4(4): 409–425, 1981

[17] Raad I. Issa. Solution of the implicitly discretised fluid flow equations by operator-splitting. *Journal of Computational Physics*, 62(1):40–65, 1986

OpenFOAM

The above methodology to simulate two-phase flow is already implemented in the software OpenFoam.[18] The solvers interFoam and multiphaseFoam were used;[19] their discretization scheme is a finite-volume approach. Convective fluxes were computed with a TVD scheme. For time-accurate computations, a first-order implicit scheme is used. Arbitrary polyhedral cells form the basis of the unstructured meshes that are required by the solvers. The subsequent droplet impact computations draw on 2D axisymmetric meshes with rectangular cell faces, 1000 × 1000 in number.

[18] Hrvoje Jasak. *Error analysis and estimation for finite volume method with applications to fluid flow.* PhD thesis, University of London, 1996; and Henrik Rusche. *Computational fluid dynamics of dispersed two-phase flows at high phase fractions.* PhD thesis, University of London, 2002

[19] Suraj S. Deshpande, Lakshman Anumolu, and Mario F. Trujillo. Evaluating the performance of the two-phase flow solver interFoam. *Computational Science & Discovery*, 5(1):014016, 2012

6
Drop impact on dry surfaces

Since the pioneering works of WORTHINGTON in the late 19th century, drop impact has attracted the attention of many researchers. The topic is relevant in both natural and industrial processes. Rain erosion, aerosol generation, spray painting and coating, fuel injection, and soldering are only a few examples. Even more complex are the accompanying boundary conditions. The droplet sphericity, impact angle, liquid and gas mediums, surface elasticity, roughness, and wetting properties constitute a myriad of variants. For **aircraft icing**, the study of drop impact at high Weber and Reynolds numbers is of primary interest. The following chapter describes the phenomenology of *dry* impact and analyses the spreading and splashing behaviours. These phenomena have a crucial influence on ice accretion and ice roughness formation. The temperature is restricted to ambient conditions, as explained in Chapter 4.4. Drop impact on thin films is discussed in the next chapter.

6.1 *Phenomenology and time evolution of liquid impact*

Taking advantage of recent possibilities in digital image recording,[1] the high-velocity drop impact is experimentally investigated using the rotating rig described on page 45. Multiple test cases, some of them already published, were selected for the analysis in this chapter, as summarized in Table 6.1. Both water and ethanol were used for the droplet fluid, enabling the derivation of a systematic variation between the Reynolds and Weber numbers.

[1] Sigurdur T. Thoroddsen, Takeharu G. Etoh, and Kohsei Takehara. High-speed imaging of drops and bubbles. *Annual Review of Fluid Mechanics*, 40:257–285, 2008

HIGH-RESOLUTION SHADOWGRAPH IMAGES of test case #D3 at a Weber number of 10,000 are depicted in Figure 6.1. The visualiza-

Reference	Case ID #	Liquid	Gas	w_{imp} $\mathrm{m\,s^{-1}}$	D mm	Re	We	St $\cdot 10^{-6}$
FASSMANN	D1	water	air	11.64	2.68	33540	5028	0.55
et al. (2013)	D2	water	air	16.18	2.70	43050	9645	0.39
BURZYNSKI and								
BANSMER (2016)	D3	water	air	14.35	3.54	50600	10000	0.34
private data	D4	ethanol	air	6.92	2.98	13135	5000	1.12
of BURZYNSKI	D5	ethanol	air	9.78	2.98	18576	10000	0.79

Table 6.1: Boundary conditions of the test cases for droplet impact on dry glass substrates that are studied in this chapter.

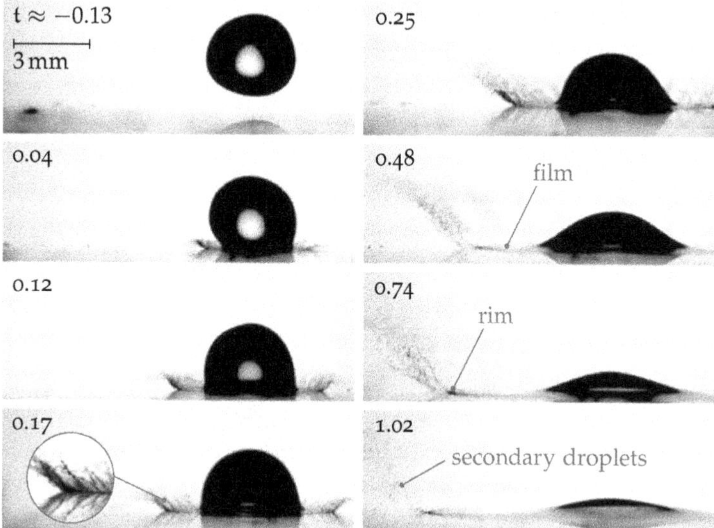

Figure 6.1: Impact of a water droplet on a dry glass substrate at ambient conditions. Case #D3, $D = 3.54\,\text{mm}$, $w_{\text{imp}} = 14.35\,\text{m}\,\text{s}^{-1}$, $\text{We} = 10\,000$, $\text{Re} = 50\,600$, $\text{St} = 3.4 \times 10^{-7}$. Shadowgraph images taken with a PCO4000 camera. Each image of the sequence corresponds to a different impact event.

tion is not time-resolved; each image of the sequence corresponds to a different impact event. A non-dimensional time is indicated in the upper left corner, given by $\mathrm{t} = t \cdot w_{\text{imp}}/D$. At $\mathrm{t} = 0$, the primary droplet virtually touches the surface. Shortly after the impact, a thin liquid sheet emerges with a very high – yet subsonic – velocity. Depending on the roughness and the chemical properties of the substrate, the surface is increasingly wetted by the droplet fluid.[2] At an early stage ($\mathrm{t} = 0.17$), the outer liquid sheet collapses into jets that are deflected upwards. Advancing in time, a thin film becomes visible and radially spreads over the surface. Its outer rim, which is thicker than the film, ejects secondary droplets that are up to two orders of magnitude smaller than the size of the primary droplet. This phenomenon is called splashing.

[2] Christophe Josserand and Sigurdur T. Thoroddsen. Drop impact on a solid surface. *Annual Review of Fluid Mechanics*, 48:365–391, 2016

6.2 *Early contact: gliding on air?*

During the early phase of droplet impact, very fast processes are triggered that alter the liquid/gas interface. Consider, therefore, the hypothetical situation of a spherical primary droplet that collides with and penetrates a flat surface with a velocity w_{imp}; see Figure 6.2. The contact radius $r(t)$ and its change over time can be estimated by

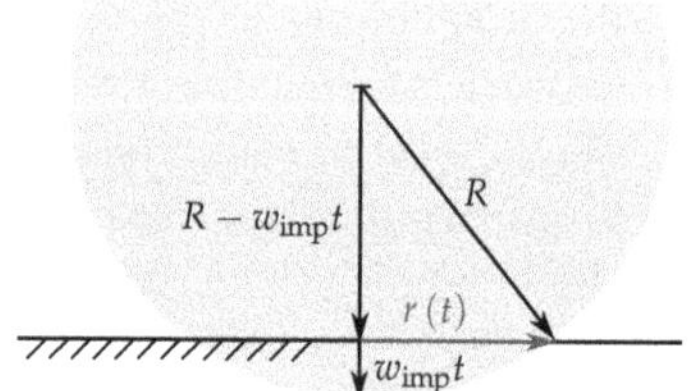

Figure 6.2: Geometric simplification for the initial phase of drop impact on a solid wall.

$$\begin{aligned} r(t) &= \sqrt{R^2 - \left(R - w_{\text{imp}} t\right)^2} \\ \frac{\partial r}{\partial t} &= \frac{2}{r} w_{\text{imp}} \left(R - w_{\text{imp}} t\right) \quad . \end{aligned}$$

At the limit of $t \to 0$, the contact radius spreads with infinite velocity

$$\lim_{t \to 0} \frac{\partial r}{\partial t} = +\infty \quad .$$

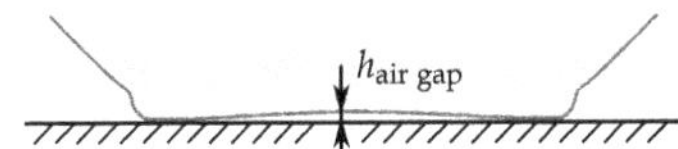

Figure 6.3: Before contacting the surface, the drop deforms and entraps a thin layer of air with a height of $h_{air\ gap}$.

THIS CONTACT SINGULARITY does not occur in reality because the gas between the droplet and the solid substrate cushions the impact.[3] Indeed, the rapidly increasing gas pressure tends to deform the liquid/gas interface by displacing the liquid to the sides. The theoretical model suggests a tendency of the liquid to further entrap the gas in small pockets; see Figure 6.3. Using interferometric measurements of droplet impacts at low Weber numbers,[4] the presence of a thin air layer was substantiated.

[3] Peter D. Hicks and Richard Purvis. Air cushioning and bubble entrapment in three-dimensional droplet impacts. *Journal of Fluid Mechanics*, 649:135–163, 2010

[4] Michelle M. Driscoll and Sidney R. Nagel. Ultrafast interference imaging of air in splashing dynamics. *Physical Review Letters*, 107(15):154502, 2011

THE HEIGHT of the entrapped air layer $h_{air\ gap}$ was recently estimated for a broader range of Stokes numbers;[5] see Figure 6.4. Herein, the parameter $\eth$, given by

$$\eth = \frac{\sigma}{\mu_{gas} \cdot w_{imp}} \mathrm{St}^{\frac{4}{3}} \quad ,$$

[5] Shreyas Mandre and Michael P. Brenner. The mechanism of a splash on a dry solid surface. *Journal of Fluid Mechanics*, 690:148–172, 2012

denotes the strength of the capillary pressure relative to the lubrication pressure. For large ratios of St / $\eth$, the droplet impact dynamics are governed by non-linear inertia. The exemplary cases #D3 and #D5 are delineated in the plot, where both water and ethanol drops surrounded by air are impacting at a Weber number of 10,000. In these cases, the height of the air layer is presumed to be $h_{air\ gap}$ = 0.3 nm. Using the approximation of a hydraulic shock, the magnitude of the air pressure can be anticipated to be approximately $p = \varrho_{liq} a_{liq} w_{imp} \approx 10^7$ Pa. Even pressures up to 10^9 Pa are predicted by the model of MANDRE and BRENNER (2012). The molecular mean free path of air at these conditions has a similar length scale as $h_{air\ gap}$, yielding a non-continuum flow with Knudsen numbers of approximately 1. Some researchers have therefore questioned the significance of the entrapped air layer.

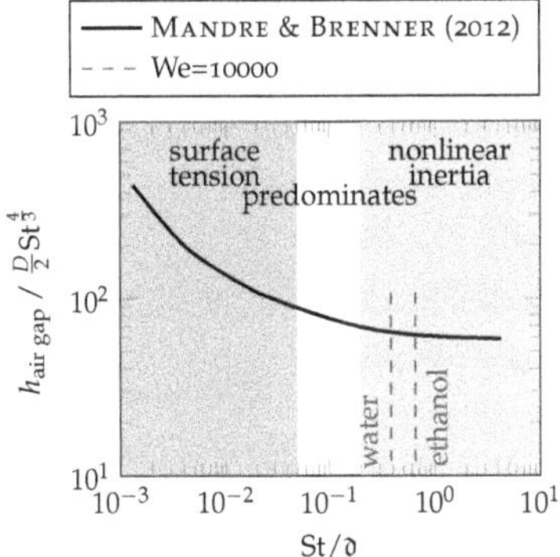

Figure 6.4: Height of the air gap as a function of St/$\eth$. Reproduced from MANDRE and BRENNER (2012) with permission.

6.3 *Spreading dynamics*

Shortly after the drop impacts the substrate, a thin film expands radially on the surface. Its boundary layer thickness δ_{film} is significantly smaller than its height h_{film}.[6] Due to capillary forces, the outer rim of the film can be approximated by a cylindrical shape;[7] see Figure 6.5. The spreading velocity of the rim is designated u_{rim}, and its volume is V_{rim}. To study the spreading dynamics of the rim, a mass and mo-

[6] Ilia V. Roisman. Inertia dominated drop collisions. II. An analytical solution of the Navier–Stokes equations for a spreading viscous film. *Physics of Fluids*, 21(5):052104, 2009

[7] Geoffrey Taylor. The dynamics of thin sheets of fluid. II. Waves on fluid sheets. *Proceedings of the Royal Society of London. Series A, Mathematical and Physical Sciences*, pages 296–312, 1959

mentum balance

$$\frac{\mathrm{d}}{\mathrm{d}t}\int_{CV(t)} \varrho_\mathrm{d} \mathrm{d}V + \int_{CS(t)} \varrho_\mathrm{d}\left(\vec{u}^m - \vec{u}^m_\mathrm{CS}\right)\cdot\vec{n}\mathrm{d}S = 0$$

$$\frac{\mathrm{d}}{\mathrm{d}t}\int_{CV(t)} \varrho_\mathrm{d}\vec{u}^m \mathrm{d}V + \int_{CS(t)} \varrho_\mathrm{d}\vec{u}^m\left(\vec{u}^m - \vec{u}^m_\mathrm{CS}\right)\cdot\vec{n}\mathrm{d}S + \cdots$$

$$\cdots + \int_{CV(t)} \varrho_\mathrm{d}\vec{a}_\mathrm{CV}\mathrm{d}V = \sum\vec{F}$$

is applied on the dash-dotted non-inertial control volume CV,[8] which deforms over time with a velocity $\vec{u}_\mathrm{CS}$ at its outer boundary and accelerating with $\vec{a}_\mathrm{CV}$. $\vec{n}$ represents the unit normal vector of the control surface pointing outwards. The fluid is incompressible. Note that the rim volume and its surface $CS(t)$ can change only due to the velocity difference between the film and rim that creates a mass flux at the left inlet CS_inlet of the control volume. This inlet is assumed to be non-deforming; $\vec{u}^m_\mathrm{CS} = 0$. The approach yields a simplification for the mass balance

[8] For more information on balance equations with deforming control volumes in a non-inertial frame of reference see the textbook of WHITE (2016). The superscript m in the velocity vectors indicates the moving frame of reference, hence $\vec{u}^m = \vec{u} - \vec{u}_\mathrm{rim}$.

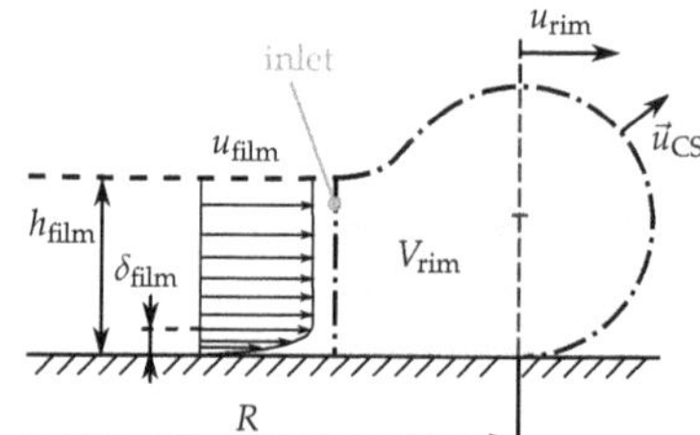

Figure 6.5: Sketch on the rim of a spreading liquid film generated by an impacting drop.

$$\varrho_\mathrm{d}\frac{\mathrm{d}V_\mathrm{rim}}{\mathrm{d}t} + \varrho_\mathrm{d}\int\limits_{CS_\mathrm{inlet}(t)} \left[\begin{pmatrix}u_\mathrm{film}\\0\end{pmatrix} - \begin{pmatrix}u_\mathrm{rim}\\0\end{pmatrix} - \begin{pmatrix}0\\0\end{pmatrix}\right]\cdot\begin{pmatrix}-1\\0\end{pmatrix}\mathrm{d}S = 0$$

$$\frac{\mathrm{d}V_\mathrm{rim}}{\mathrm{d}t} - h_\mathrm{film}\cdot 2\pi R\left(u_\mathrm{film} - u_\mathrm{rim}\right) = 0 \quad ,$$

while neglecting the boundary layer effect of the film flow and approximating the radial inlet boundary surface as $h_\mathrm{film}\cdot 2\pi R$. Further assuming a predominantly uniform velocity field inside the control volume $\vec{u}^m = \vec{0}$, the momentum balance in the radial direction can be written as

$$\frac{\mathrm{d}}{\mathrm{d}t}\int\limits_{CV(t)}\vec{0}\,\mathrm{d}V + \int\limits_{CS_\mathrm{inlet}(t)}\left(u_\mathrm{film} - u_\mathrm{rim}\right)\left[\left(\vec{u}_\mathrm{film} - \vec{u}_\mathrm{rim} - \vec{0}\right)\cdot\vec{n}\right]\mathrm{d}S + \cdots$$

$$\cdots + \int\limits_{CV(t)}\frac{\partial u_\mathrm{rim}}{\partial t}\mathrm{d}V = \frac{1}{\varrho_\mathrm{d}}\sum F_r$$

$$-h_\mathrm{film}\cdot 2\pi R\left(u_\mathrm{film} - u_\mathrm{rim}\right)^2 + V_\mathrm{rim}\frac{\partial u_\mathrm{rim}}{\partial t} = \frac{F_{\sigma,r} + F_{\mathrm{visc},r}}{\varrho_\mathrm{d}} \quad .$$

The right-hand side is composed of surface tension forces $F_{\sigma,r}$ and viscous forces $F_{\mathrm{visc},r}$. The latter are caused by the wall shear stress τ_w, which can be approximated using the average of the film and rim velocities and the boundary layer thickness δ_film. The viscous length scale l_visc corresponds to a fraction l_rel of the rim diameter.

This approach yields

$$F_{\sigma,r} + F_{\text{visc},r} = -2\pi R\,(2\sigma + \tau_{\text{w}} \cdot l_{\text{visc}})$$
$$\approx -2\pi R \left(2\sigma + \mu_{\text{d}} \frac{u_{\text{rim}} + u_{\text{film}}}{2\delta_{\text{film}}} \cdot l_{\text{rel}} \sqrt{\frac{V_{\text{rim}}}{2\pi R}} \right) \quad .$$

Furthermore, the film height h_{film}, film velocity u_{film} and boundary layer thickness δ_{film} can be described by a remote asymptotic solution:[9]

$$h_{\text{film}} = D \cdot \left(\frac{0.05}{\mathrm{t}^2} + 0.48 \sqrt{\frac{\mathrm{t}}{\text{Re}}} \right)$$
$$u_{\text{film}} = w_{\text{imp}} \cdot \frac{1}{\mathrm{t} + 0.25} \qquad (6.1)$$
$$\delta_{\text{film}} = 1.88D \cdot \sqrt{\frac{\mathrm{t}}{\text{Re}}} \quad ,$$

where $\mathrm{t} = t \cdot w_{\text{imp}}/D$ represents the non-dimensional time.

[9] Alexander L. Yarin. *Free liquid jets and films*. Interaction of Mechanics and Mathematics Series. Longman Scientific & Technical, Harlow, Essex, 1993. ISBN 0582102952

THE RESULTING SYSTEM of ordinary differential equations is solved for u_{rim} with the constraint of $\mathrm{t} > 1$ to ensure a fully developed film flow. A plot of the rim velocity is shown in Figure 6.6. The film and rim velocities are initially on the same order of magnitude. Viscous and capillary forces decelerate the rim. At $\mathrm{t} = 6$, the rim is at rest but still growing in size due to the remaining film flow. The theory is supported by experimental data and interFoam simulations.

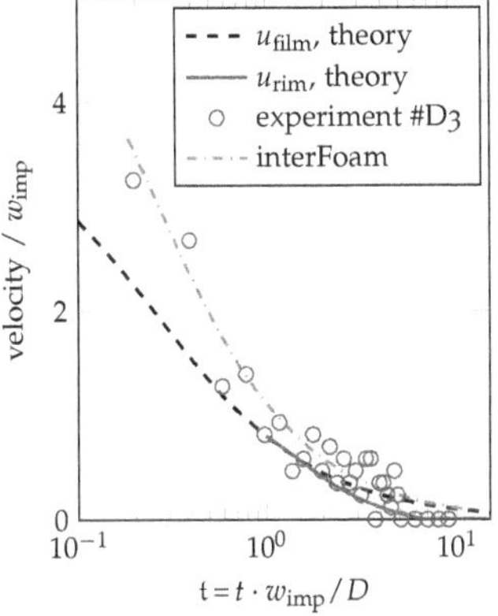

Figure 6.6: Velocity of the spreading rim and its accruing film. Comparison of the data from BURZYNSKI and BANSMER (2016) at We=10,000 with the developed theory and interFoam computations.

THE TEMPORAL EVOLUTION OF THE SPREADING RADIUS of the droplet is shown in Figure 6.7. The initial slope corresponds well to the integrated film velocity u_{film} of the remote asymptotic solution (dashed line); see again equation 6.1. When integrating the rim velocity u_{rim} of the above theory, the deceleration of the rim is visible. However, the predicted maximum spreading radius does not match the experimental data as well as the empirical model from ROISMAN (2009).

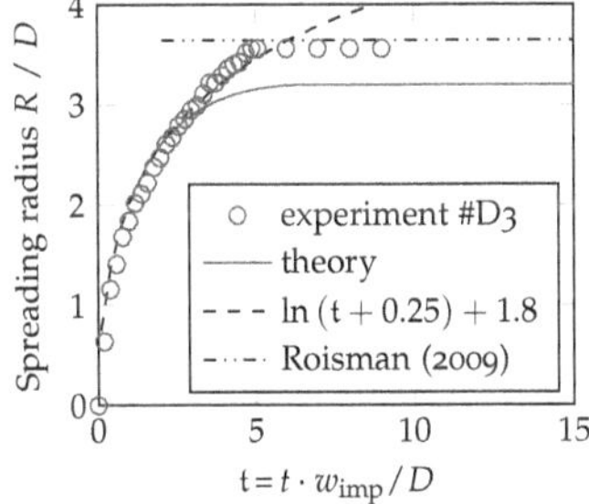

Figure 6.7: Comparison of the data from BURZYNSKI and BANSMER (2016) with the developed theory and the empirical correlation of ROISMAN (2009) on the maximum spreading radius.

6.4 *Lamella levitation*

The viscosity of the drop strongly influences the splashing dynamics. Therefore, experiments have been performed using ethanol, whose dynamic viscosity μ_{liq} is approximately 25% higher than that of water. Figure 6.8 shows shadowgraph images of such ethanol drops impacting at a Weber number of 10,000. A comparison of these images with the image sequence of impinging water droplets in Figure 6.1 reveals a different behaviour of the spreading film. The film forms a

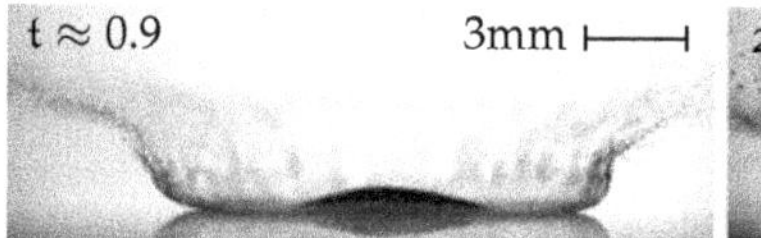

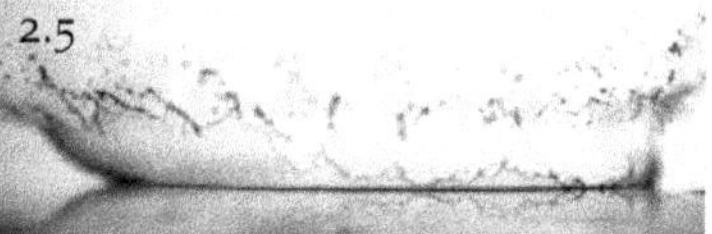

Figure 6.8: For an ethanol drop impact at We=10,000 (test case #D5), the lamella detaches from the substrate surface. The splashing mechanism is therefore different from that of water.

liquid lamella that lifts off from the substrate and creates a free liquid sheet. The outer rim thereafter destabilizes and forms cusps, from which thin liquid jets emerge that ultimately break down to small secondary droplets.

To EXPLAIN the phenomenon of a levitating lamella, two mechanisms are contemplated.[10] On the one hand, this is a rapidly moving outer rim of a spreading film with a large advancing contact angle that incorporates a thin layer of air beneath it. On the other hand, a large vortex was identified above the spreading film[11] that creates a suction force deflecting the lamella.

[10] Henry Vu, Darren Banks, and Guillermo Aguilar. Examining viscosity and surface wettability on lamella lift dynamics and droplet splashing. *Atomization and Sprays*, 21(4), 2011

[11] Irmgard Bischofberger, Kelly W. Mauser, and Sidney R. Nagel. Seeing the invisible – Air vortices around a splashing drop. *Physics of Fluids*, 25(9): 091110, 2013

6.5 *Secondary droplets*

During splashing, secondary droplets are generated. In the case of aircraft icing, these droplets are transported further downstream by the surrounding airflow, where they might reimpinge on exposed aircraft components.[12] Additionally, the deposited mass on the substrate is smaller than that of the primary droplet, which alters the mass balance of ice accretion; see Chapter 9.4.

[12] David R. Bilodeau, Wagdi G. Habashi, Marco Fossati, and Guido S. Baruzzi. Eulerian modeling of supercooled large droplet splashing and bouncing. *Journal of Aircraft*, 52(5):1611–1624, 2015

SIZING THE SECONDARY DROPLETS reveals interesting information on the splashing process. For this purpose, an ensemble of more than one thousand time-stamped, high-resolution shadowgraph images was evaluated with an algorithm that incorporates a depth-of-field correction.[13] The results for test cases #D1 and #D2 of impinging water droplets are depicted in Figure 6.9 (left), where the mean diameter of the secondary droplets $\langle d_{\text{sec}} \rangle$ is plotted over time. Remarkably, the secondary droplets are very small, with a diameter only approximately 2% of the primary droplet diameter D. The higher the Reynolds number, the smaller the secondary droplets. It is most likely that the increasing surplus of impact kinetic energy promotes the enhanced formation of free surfaces.

[13] Ki Seong Kim and Sung-Soo Kim. Drop sizing and depth-of-field correction in TV imaging. *Atomization and Sprays*, 4(1), 1994

ADVANCING IN TIME, the generated secondary droplets become larger; however, their further growth is suddenly attenuated at $t \approx 7$. Since the secondary droplets are emerging from the outer rim of either the film or levitated lamella, the developed theory of Chapter 6.3

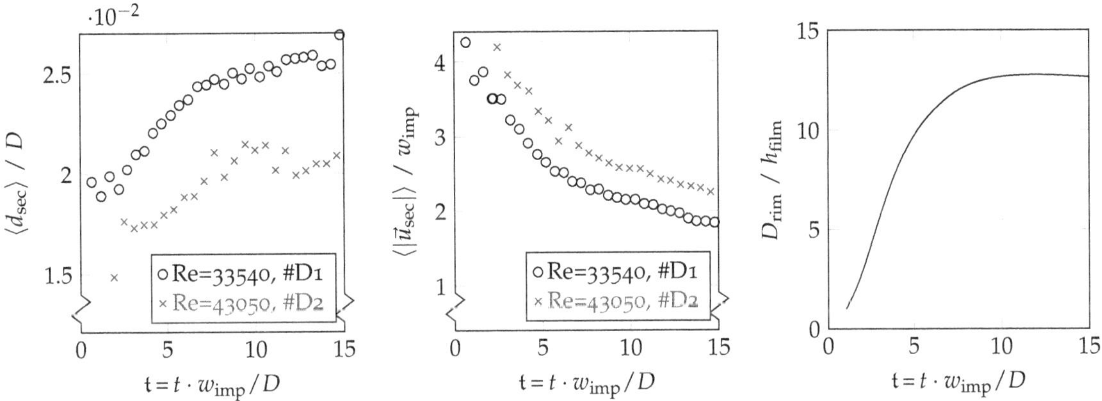

Figure 6.9: Secondary droplets that change their dynamics at $\mathrm{t} = 8$. *Left:* The mean diameter of the secondary droplets increases over time. *Centre:* The mean velocity of the secondary droplets decreases over time. Data from FASSMANN et al. (2013). *Right:* The theory presented in Section 6.3 predicts a growing ratio between the rim size D_{rim} and film thickness h_{film} over time.

might partially explain their arrested growth. Figure 6.9 (right) thus illustrates the ratio between the rim diameter $D_{\mathrm{rim}} = \sqrt{\frac{V_{\mathrm{rim}}}{2\pi R}}$ and film height h_{film} over time. This ratio constantly grows for $\mathrm{t} < 4$, and the slope decreases thereafter until reaching a nearly constant level at $\mathrm{t} = 9$. In the transient phase $4 < \mathrm{t} < 9$, the droplet spreading terminates. Hence, the momentum balance of the film flow at the outer rim and secondary droplets is affected, thus diminishing the generation of free surfaces for secondary droplets.

THE MEAN VELOCITY of the secondary droplets $\langle|\vec{u}_{\mathrm{sec}}|\rangle$ can be estimated from two consecutive shadowgraph images with a sufficiently small time difference. Figure 6.9 (centre) highlights a decreasing velocity over time for the aforementioned test cases. $\langle|\vec{u}_{\mathrm{sec}}|\rangle$ is approximately three times higher than the impact velocity w_{imp}. Similar to the droplet sizing mechanism, a higher kinetic energy of impact creates faster secondary droplets. A changing behaviour of $\langle|\vec{u}_{\mathrm{sec}}|\rangle$ at $\mathrm{t} \approx 7$ is similarly observed and further substantiates the influence of the cessation of the spreading process.

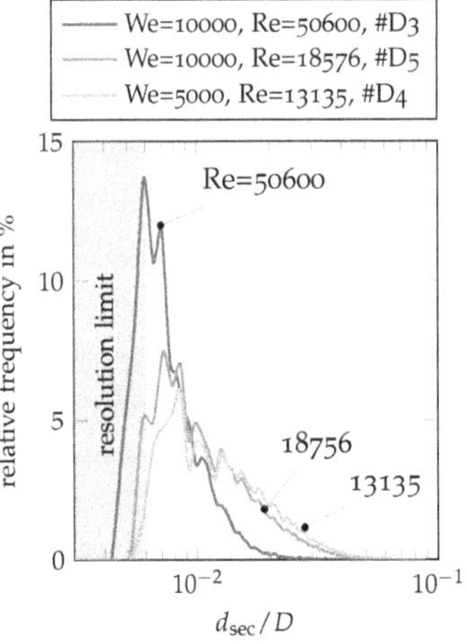

Figure 6.10: Size distribution of secondary droplets for varying Reynolds and Weber numbers. Private data from BURZYNSKI.

OF THE REYNOLDS NUMBER OR THE WEBER NUMBER, which is the more suitable non-dimensional parameter for quantifying the outcome of the secondary droplets? For a systematic variation, experiments with ethanol drops were conducted and compared with the impact results of water drops. Figure 6.10 depicts the size distribution of secondary droplets for test cases #D3-#D5. When impacting at a Weber number of 10,000, the ethanol drop creates significantly larger secondary droplets relative to water. This effect is also clearly visible when contrasting the shadowgraph images presented in Figure 6.1 and 6.8. Notably, the Reynolds number is well correlated with

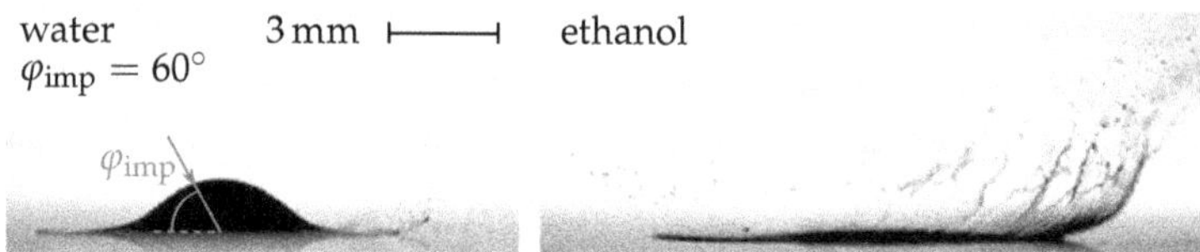

Figure 6.11: Oblique drop impact on a dry surface at We=5000. Images courtesy of LOSTER and BURZYNSKI.

the size of the secondary droplets; a larger Reynolds number corresponds to smaller ejected droplets.

6.6 *Oblique impact*

Previously, the normal drop impact was considered. When the impact angle φ_{imp} is less than 90°, a tangential impact velocity is superimposed over the subsequent phenomena.[14] One hemisphere of the impacting drop thus involves an increased kinetic energy, inducing an asymmetric splashing process. Figure 6.11 illustrates this asymmetry by contrasting the impacts of water and ethanol drops at We=5000. The water drop splashes on its right-hand side, but no secondary droplets are visible on the left-hand side. The asymmetry is even more prominent with the ethanol drop, whose lamella height is constantly increasing in size farther to the right. Since many technical applications involve oblique impact phenomena, this investigation is of great importance for future research.

[14] Damon G. K. Aboud and Anne-Marie Kietzig. Splashing threshold of oblique droplet impacts on surfaces of various wettability. *Langmuir*, 31(36): 10100–10111, 2015

7 Drop impact on liquid films

During glaze ice accretion on an aircraft, thin liquid films of water can be found lying on the top of the deposited ice. Drop impact on liquid films is hence a strongly influential process for the entire course of icing. The drop and film interact throughout the impact sequence; thus, the mechanisms of secondary droplet production are different from those of dry impact. Considering that only 5% of the primary droplet volume is converted to secondary droplets during dry impact, it is notable that more than 100% can be generated during impact on liquid films.

7.1 Phenomenology

The drop impact on liquid films was experimentally investigated using the rotating rig described on page 45. The impact phenomena at high Weber numbers are again the focus of the studies. The presence of a liquid film necessitates further boundary conditions, given by the film height h_f and the average film velocity $\bar{u}_f$. Multiple test cases have been selected for the analysis in this chapter; see the summary in Table 7.1. The choice of the non-dimensional film height h_f/D follows those conditions of an aerofoil exposed to a water spray.[1] Furthermore, the ratio between the film and impact velocity is typically very small for icing conditions, $\bar{u}_f/w_{imp} < 1\,\%$.[2]

[1] Earl C. Hastings Jr and Leonard M. Weinstein. Preliminary indications of water film distribution and thickness on an airfoil in a water spray. Technical Report NASA-TM-85796, NASA Langley Research Center, 1984

[2] Yves Bourgault, Héloïse Beaugendre, and Wagdi G. Habashi. Development of a shallow-water icing model in FENSAP-ICE. *Journal of Aircraft*, 37(4): 640–646, 2000

Reference	Case ID #	D mm	w_{imp} m s^{-1}	h_f/D	$\bar{u}_f/w_{imp}$	Re	We
Fassmann (2015)	W1a	2.66	9.58	8.6%	22%	27180	3376
|	W1b	2.68	9.58	16.0%	20%	29240	3412
|	W2	2.68	11.57	14.2%	11%	31860	4970
Burzynski &	W3a	3.00	7.44	5%	5%	20596	2281
Bansmer (2018)	W3b	3.00	7.44	13%	15%	20596	2281
|	W3c	3.00	7.44	14%	29%	20596	2281
|	W4a	3.00	10.53	3%	4%	29113	4569
|	W4b	3.00	10.53	4%	10%	29113	4569
|	W4c	3.00	10.53	5%	20%	29113	4569
computation	W5	3.00	7.44	3.3%	0	22253	2283

Table 7.1: Boundary conditions of the test cases for droplet impact on thin films that are studied in this chapter. Water was used for the drop liquid, and air was the surrounding gas.

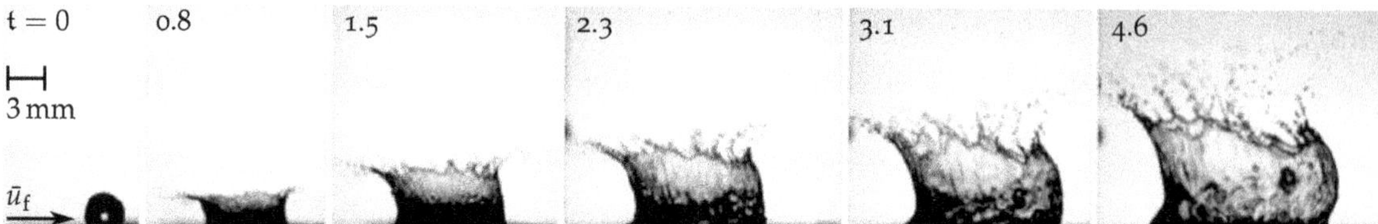

Figure 7.1: After a drop impacts on a liquid film, a crown emerges. The film velocity deflects the crown. At the upper rim, secondary droplets are generated. Test case #W3b.

TIME-RESOLVED shadowgraph images of test case #W3b – a water drop impact at We=2281 – are depicted in Figure 7.1. The time t was non-dimensionalized to $\mathrm{t} = t \cdot w_{\mathrm{imp}}/D$ and is annotated at the upper left corner of each shadowgraph. Shortly after the drop touches the film, a liquid column rises nearly perpendicularly from the surface. Advancing in time, the column spreads, and its upper free rim becomes unstable. Small cusps are formed, which is why the term *crown* is typically employed to describe the liquid column. At the tip of each cusp, a thin liquid jet emanates and subsequently disintegrates into small secondary droplets. Due to the film velocity $\bar{u}_f$, the crown deforms at the later stages.

7.2 *Computations reveal the film behaviour*

The computational scheme presented on page 45 can be extended to simulate the flow of three immiscible fluids, i.e., the droplet, film and surrounding gas. For this purpose, equation (5.3) is complemented by the density and viscosity of the third phase, and a second scalar indicator of type χ for the VOF interface modelling is introduced.

THEREUPON, A TWO-DIMENSIONAL AXISYMMETRIC MESH of 1.5×10^6 cells is used to simulate test case #W5, which represents a water drop impact at We=2283 on a film at rest. The goal is to emphasize the interplay between the drop and film during early impact, which is very challenging to realize experimentally. Figure 7.2 illustrates the temporal evolution of the three phases using different shades of blue. In the white regions close to the interfaces, the individual fractional vol-

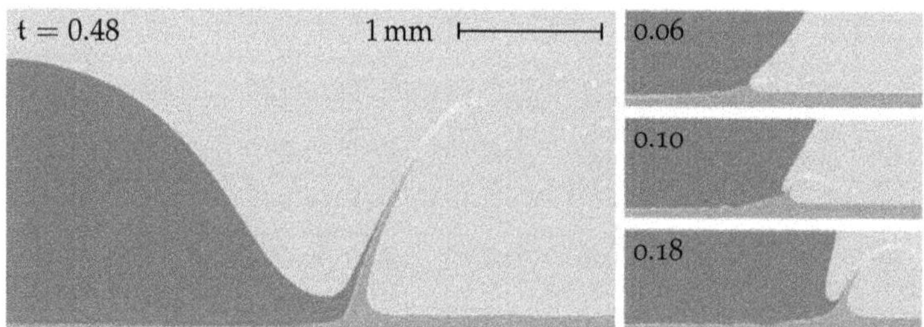

Figure 7.2: Three-phase simulation of test case #W5. The drop displaces the liquid film and causes the formation of a thin liquid sheet, which forms the crown at a later stage. The VOF scalar indicators for each phase are plotted in the range $0.1 < \chi \leq 1$. At locations where $\chi < 0.1$, the colouring is white. Data provided by DAVID BURZYNSKI.

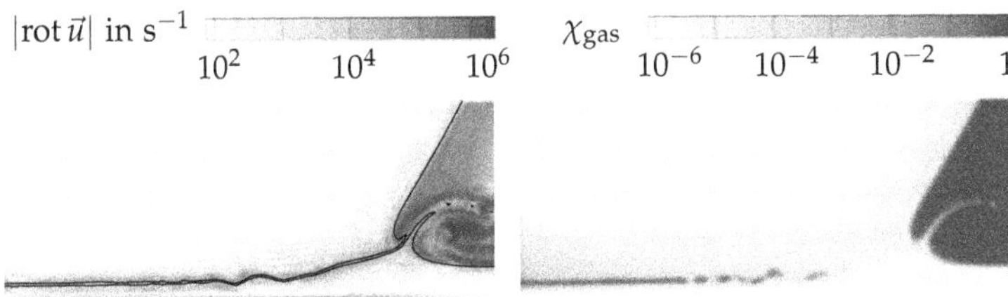

Figure 7.3: Three-phase simulation of test case #W5. Vorticity *(left)* and fractional volume of gas *(right)* at the interface between the droplet and film for $t = 0.10$. Data provided by DAVID BURZYNSKI.

umes χ of the droplet, film and gas are smaller than 10 %. Shortly after impact, at $t = 0.06$, the cohesive forces between the drop and film promote a partial reconnection of the interfaces, and a thin liquid sheet is radially ejected. Due to the droplet's high inertia, the underlying film is radially displaced. However, the surrounding film at rest counteracts the spreading momentum. Consequently, the mass flux is redirected upwards, away from the substrate. Thereafter, the base of the arising crown acts as a kinematic discontinuity.[3] The droplet fluid cannot penetrate through the mounting crown (at $t = 0.18$) and is similarly deflected upwards. Hence, the crown consists of two fluids: that of the droplet and that of the film.

[3] Alexander L. Yarin and Daniel A. Weiss. Impact of drops on solid surfaces: self-similar capillary waves, and splashing as a new type of kinematic discontinuity. *Journal of Fluid Mechanics*, 283:141–173, 1995

SMALL INSTABILITIES at the interface between the droplet and film are visible at $t = 0.10$. Such instabilities are caused by the surrounding gas that is entrapped during the early stages of impact. Figure 7.3 (right) depicts the fractional volume of gas χ_{gas} in each computational cell. Dark blue regions indicate the locations of air. Light blue regions are an effect of the small numerical diffusion; note that χ_{gas} is on the order of 10^{-5} in these regions. Notably, a thin layer of air, partially collapsed into small bubbles, can be found between the droplet and film. This condition generates an increased vorticity in the flow field; see the left part of Figure 7.3. Indeed, a von Kármán vortex street was recently identified in this region for low Reynolds numbers[4] and is hypothesized to be another instability mechanism that causes droplet splashing.[5]

[4] Alfonso A. Castrejón-Pita, Rafael Castrejón-Pita, and Ian M. Hutchings. Experimental observation of von Kármán vortices during drop impact. *Physical Review E*, 86(4):045301, 2012

[5] Marie-Jean Thoraval, Kohsei Takehara, Takeharu Goji Etoh, Stéphane Popinet, Pascal Ray, Christophe Josserand, Stéphane Zaleski, and Sigurdur T. Thoroddsen. von Kármán vortex street within an impacting drop. *Physical Review Letters*, 108(26):264506, 2012

7.3 *Influence of the film velocity*

For $\mathrm{We} = 2281$, the effect of a varying film velocity $\bar{u}_f$ was studied. Figure 7.4 (left) shows the corresponding shadowgraph images of the crown at $t = 4.6$. For low film velocities, $\bar{u}_f / w_{imp} = 5\,\%$, a nearly axisymmetric shape of the crown is observed. With increasing film velocity, the lamella curves to the left, i.e., in the opposite direction of the film velocity. A momentum balance at the base of the crown partially explains this asymmetry: depending on the bearing of the spreading fluid, the fluid is either competing against or augmenting

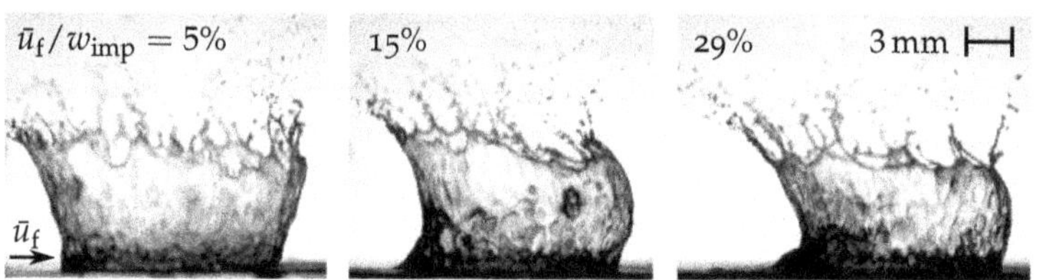

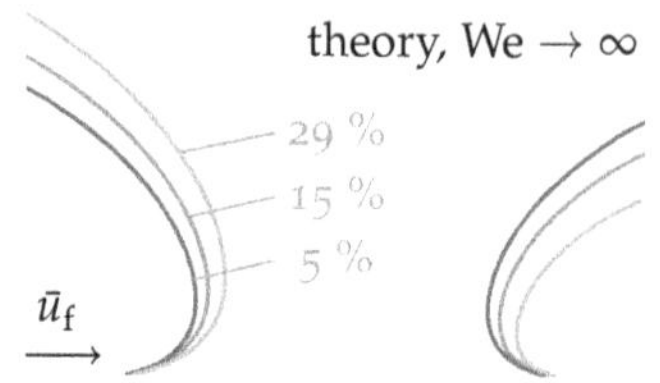

Figure 7.4: Effect of the varying film velocity. Crown shape at $\mathfrak{t} = 4.6$. *Left:* Shadowgraph images for test cases #W3a-c. *Right:* Theory of Roisman and Tropea (2002) for We $\rightarrow \infty$, $h_f/D = 13\,\%$, varying $\bar{u}_f/w_{imp}$.

the movement of the outer film. Modelling[6] this momentum balance for We $\rightarrow \infty$ yields a crown shape, which is plotted in the right part of Figure 7.4. Note that neither the gas flow nor viscous effects are included in the model. An increased height of the left flank of the crown is correctly predicted, but the curvature of the right flank does not coincide with the measurements.

[6] Ilia V. Roisman and Cameron Tropea. Impact of a drop onto a wetted wall: description of crown formation and propagation. *Journal of Fluid Mechanics*, 472:373–397, 2002

7.4 *Crown dimensions*

Time-resolved shadowgraph images of the test cases #W3a-c and #W4a-c are processed to estimate the characteristic dimensions of the evolving crown geometry. Both the height $\mathfrak{H}_{cro}$ and diameter $\mathfrak{D}_{cro}$ are determined directly; see Figure 7.5. The diameter is tapped at the narrowest section of the crown. The height information is an average between the left and right flanks; its precise determination is impeded by the fuzzy position of a disintegrating crown rim. An indirect measurement, which is described in the next section, yields the thickness of the lamella $\mathfrak{T}_{cro}$. Note that all the above parameters are non-dimensionalized by the primary droplet diameter D.

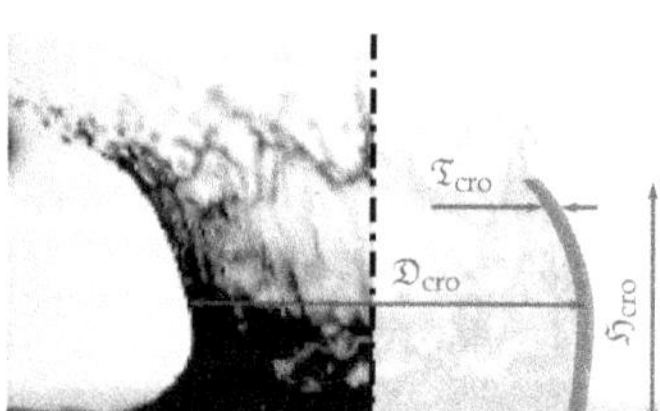

Figure 7.5: Approximate height $\mathfrak{H}_{cro}$, diameter $\mathfrak{D}_{cro}$ and thickness $\mathfrak{T}_{cro}$ of the crown that are non-dimensionalized by the primary droplet diameter D.

Plotting the experimental data of the temporal evolution of the crown diameter $\mathfrak{D}_{cro}$ provides useful information on the spreading behaviour; see Figure 7.6. Clearly, increasing the Weber number promotes greater spreading, and the high inertial forces of the impacting drop facilitate the displacement of the thin film. Furthermore, both the film speed and film thickness have been varied simultaneously in test cases #W3a-c and #W4a-c. It is the film thickness, however, that mainly influences the spreading. The thicker the film, the higher its inertia, which has to be surmounted by the spreading droplet. Generally, a square root law for the crown expansion is observed, $\mathfrak{D}_{cro} \sim \sqrt{\mathfrak{t}}$, which is also reflected by the three-phase simulation of test case #W5. The crown height and diameter typically grow proportionally over time. Therefore, the crown also piles up following a square root law: $\mathfrak{H}_{cro} \sim \sqrt{\mathfrak{t}}$.

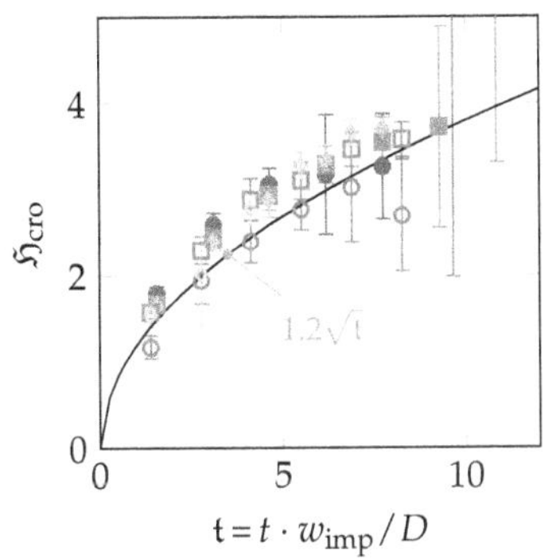

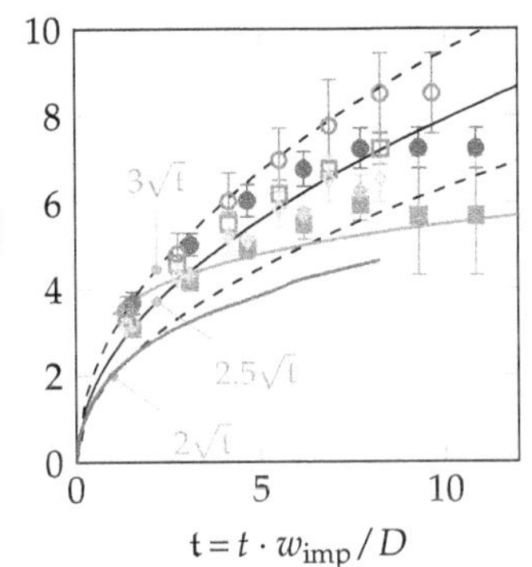

Figure 7.6: Experimental data on the height *(left)* and the spreading *(right)* of the crown for test cases #W3a-c and #W4a-c. Additional data from a three-phase simulation of test case #W5 and the theoretical model of Roisman and Tropea (2002) for We $\to \infty$ and the film properties of test case #W3b.

Since the spreading dynamics do not vary by many orders of magnitude when the impact parameters vary, appropriate scaling laws are often hard to identify by quantitative arguments alone. Numerous correlations have been elaborated in the past;[7] such correlations vary from simple relations, e.g., $\mathfrak{D}_{cro} = 2\sqrt{t}$, to complex models characterizing the balance between viscous, inertia and capillary forces.[8] Despite their different formulations, most of these relations show satisfactory agreement with experimental data; see, for instance, the aforementioned model of Roisman and Tropea (2002).

[7] Alexander L. Yarin. Drop impact dynamics: splashing, spreading, receding, bouncing... *Annual Review of Fluid Mechanics*, 38:159–192, 2006

[8] Christophe Josserand, Pascal Ray, and Stéphane Zaleski. Droplet impact on a thin liquid film: anatomy of the splash. *Journal of Fluid Mechanics*, 802:775–805, 2016

7.5 *Crown breakup*

Typically, the splash of a droplet on a thin liquid film is subdivided into four phases:[9] crown formation, instability of the free rim, jet formation, and atomization into small secondary droplets. This phenomenology is visualized in Figure 7.1 and is further discussed in the next section.

[9] Gianpietro E. Cossali, Aldo Coghe, and Marco Marengo. The impact of a single drop on a wetted solid surface. *Experiments in fluids*, 22(6):463–472, 1997

For high Weber numbers, another process of crown breakup is observed;[10] see Figure 7.7. Holes spontaneously form in the middle of the crown and spread throughout the lamella until they reach either the crown rim or its base. Occasionally, the lamella separates independently from the crown base. In both cases, the crown completely disintegrates a short time later. To find the cause of these instabilities, high-resolution shadowgraphy was performed. Indeed, small bubbles in the lamella are identified that burst after reaching the middle of the crown. When the lamella is locally thin enough, one may hypothesize that the bursting of these bubbles produces the holes. It is still unclear, however, whether the bubbles originate from an early impact stage where air is entrapped between the droplet and film.

[10] David A. Burzynski and Stephan E. Bansmer. Droplet splashing on thin moving films at high Weber numbers. *International Journal of Multiphase Flow*, 101:202–211, 2018

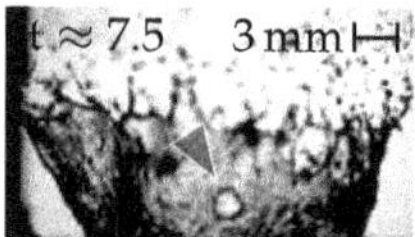

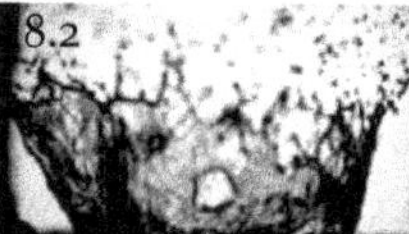

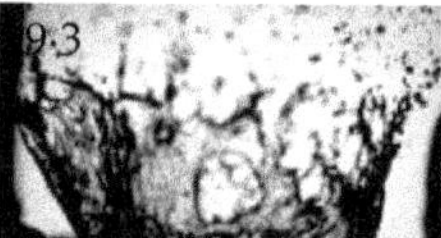

Figure 7.7: A hole grows in the midst of the crown and coalesces with a second hole that emerges at the base, ultimately initiating breakup of the entire crown. Data from BURZYNSKI and BANSMER (2018), test case #W4a at We=4600.

A SIMILAR INSTABILITY MECHANISM is observed when puncturing thin soap films.[11] A hole forms whose edge propagates with nearly constant velocity u_{hole} away from the point of puncture. An energy balance between the liberated surface energy and kinetic energy yields a relation for the film thickness. Drawing parallels by applying this paradigm, the non-dimensional crown thickness of the drop impact scenario can be calculated as

$$\mathfrak{T}_{\mathrm{cro}} = \frac{2\sigma}{\varrho_{\mathrm{liq}} D u_{\mathrm{hole}}^2} \quad .$$

In the investigated parameter space of test cases #W3a-b and #W4a-b, the crown thickness varied between 5 and 30 µm.

[11] Fred E. C. Culick. Comments on a ruptured soap film. *Journal of Applied Physics*, 31(6):1128–1129, 1960

7.6 *Secondary droplets*

The typical process of secondary droplet formation is illustrated in Figure 7.8. The thin lamella feeds the free rim, which is – owing to surface tension forces – cylindrical in shape with radius R_{rim}. In the circumferential direction, instabilities of a length scale l^*_{rim} form, resulting in cusps that contract into fingerlike jets with radius R_{fing}. Depending on their velocity, these jets disintegrate on a length scale l^*_{fing} into secondary droplets of diameter d_{sec}.

BY EVALUATING high-resolution shadowgraph images of test case #W3a, quantitative information on the above characteristic dimensions is extracted; see the left part of Figure 7.8. We determine that $2R_{\mathrm{rim}} = 275\,\mu\mathrm{m}$, $l^*_{\mathrm{rim}} = 1600\,\mu\mathrm{m}$, $2R_{\mathrm{fing}} = 230\,\mu\mathrm{m}$, $l^*_{\mathrm{fing}} = 731\,\mu\mathrm{m}$. The thickness of the lamella can be deduced from the fringe patterns of propagating fluid waves directly below the rim. Nonlinear theory[12] yields

$$D \cdot \mathfrak{T}_{\mathrm{cro}} = \frac{0.7\lambda_{\mathrm{fringe}}}{2\pi} = \frac{0.7 \cdot 75\,\mu\mathrm{m}}{2\pi} = 8.4\,\mu\mathrm{m} \quad .$$

[12] Martine Prévost and Dominique Gallez. Nonlinear rupture of thin free liquid films. *The Journal of Chemical Physics*, 84(7):4043–4048, 1986

THEORETICAL CONSIDERATIONS at the free rim[13] allow several relationships to be established between the abovementioned characteristic dimensions. The ratio $l^*_{\mathrm{rim}}/R_{\mathrm{rim}} \approx 8.9$, which results from an analysis of the growth rate of small harmonic disturbances, is of the

[13] Ilia V. Roisman, Kristijan Horvat, and Cameron Tropea. Spray impact: rim transverse instability initiating fingering and splash, and description of a secondary spray. *Physics of Fluids*, 18 (10):102104, 2006

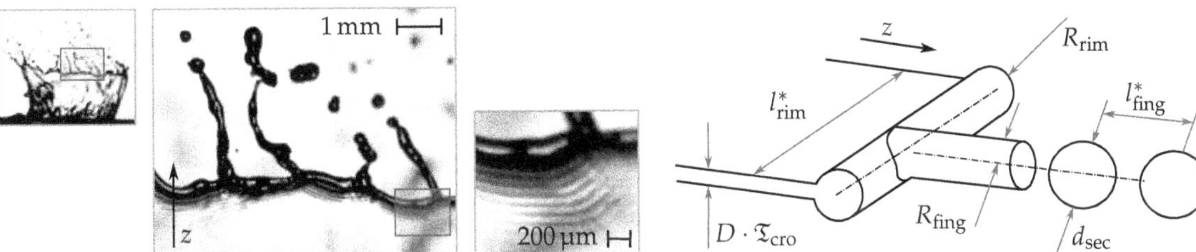

Figure 7.8: High-resolution shadowgraph image of test case #W3a emphasizing the free rim *(left)*. Schematic of the instability mechanism at the free rim including characteristic dimensions *(right)*.

same order as the length scale of a Rayleigh instability of a cylindrical jet. The same relationship holds for the ratio $l^*_{\text{fing}}/R_{\text{fing}} < 8.9$; however, the high jet velocity together with the surrounding gas reduces the length scale. Indeed, the associated aerodynamic Weber numbers suggest a transition between the Rayleigh instability and the first wind-induced breakup.[14] Moreover, the jet radius and secondary droplet size are related by the length scale l^*_{fing}. A comparison of the volumes gives

$$\begin{aligned} \frac{\pi}{6} d_{\text{sec}}^3 &= \pi R_{\text{fing}}^2 \cdot l^*_{\text{fing}} \\ &= \pi R_{\text{fing}}^3 \cdot \frac{l^*_{\text{fing}}}{R_{\text{fing}}} \\ \Rightarrow \quad \frac{d_{\text{sec}}}{R_{\text{fing}}} &\approx 3.8 \quad . \end{aligned}$$

[14] Sung P. Lin and Rolf D. Reitz. Drop and spray formation from a liquid jet. *Annual Review of Fluid Mechanics*, 30(1): 85–105, 1998

Geometric relations between the jet, rim and lamella thickness are obtained by a momentum balance at the free rim. The theoretical predictions and the experimental results are compared in Table 7.2 and show very good agreement.

THE QUANTITATIVE PROPERTIES of the secondary droplets for test case #W1b are depicted in Figure 7.9. On the left, the volume-weighted size distribution of the secondary droplets is shown. Its peak is located at $d_{\text{sec}}/D = 0.13$, and the droplets are hence much larger than those generated during a dry impact. The ratio between the droplet size and crown thickness according to the aforementioned theory is well recognized. The temporal evolution of the secondary droplets resembles the observations for dry impact; see the middle and right parts of Figure 7.9. The mean size $\langle d_{\text{sec}} \rangle$ is gradually increasing, while the velocity decreases over time. Similarly, the slope

Reference	$\frac{l^*_{\text{rim}}}{R_{\text{rim}}}$	$\frac{R_{\text{fing}}}{R_{\text{rim}}}$	$\frac{l^*_{\text{fing}}}{R_{\text{fing}}}$	$\frac{d_{\text{sec}}}{R_{\text{fing}}}$	$\frac{R_{\text{fing}}}{D\mathfrak{T}_{\text{cro}}}$	$\frac{d_{\text{sec}}}{D\mathfrak{T}_{\text{cro}}}$
theory, ROISMAN et al. (2006)	8.9	1.22	<8.9	<3.8	10.8	40.7
experiment #W3a	11.6	0.8	6.4	3.4	13.8	41.6

Table 7.2: Comparison of non-dimensional groups for the crown's rim instability.

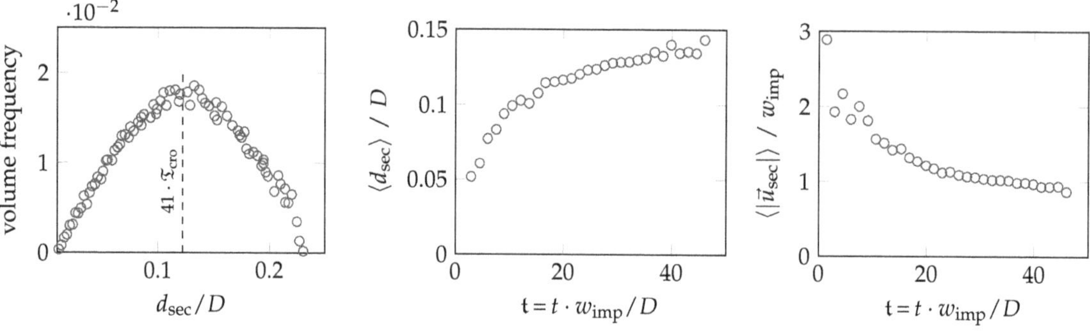

Figure 7.9: *Left:* Volume-weighted frequency distribution of secondary droplets, test case #W1b. Temporal evolution of the size *(middle)* and velocity *(right)* of the secondary droplets of test case #W1b.

changes at $t \approx 10$, which corresponds to the termination of the spreading phase.

DETERMINING the total volume of the secondary droplets is a non-trivial task. The shadowgraph images with their small field of view and the even smaller depth of field constitute only a limited measurement volume. Integration over space can be performed assuming axisymmetric conditions of secondary droplet generation. Integration over time is more complex because the droplets cross the boundaries of the measurement volume. FASSMANN (2015) included the velocity information of the secondary droplets to compute the mass flux, which ultimately yielded the total volume V_{sec}. Another approach involves the estimation of the crown volume V_{cro}. An approximation using Cavalieri's principle for the main part of the crown and a cylindrical shape of the free rim yields

$$\frac{V_{cro}}{V_d} = \frac{\mathfrak{H}_{cro} \cdot \pi\, \mathfrak{D}_{cro} \cdot \mathfrak{T}_{cro} \cdot D^3 + \pi \left(\frac{R_{rim}}{D\mathfrak{T}_{cro}}\right)^2 \mathfrak{T}_{cro}^2 \cdot \pi\, \mathfrak{D}_{cro} \cdot D^3}{\frac{\pi}{6} D^3}$$
$$= 6\, \mathfrak{D}_{cro} \mathfrak{T}_{cro} \left[\mathfrak{H}_{cro} + \pi \left(\frac{R_{rim}}{D\mathfrak{T}_{cro}}\right)^2 \mathfrak{T}_{cro} \right] .$$

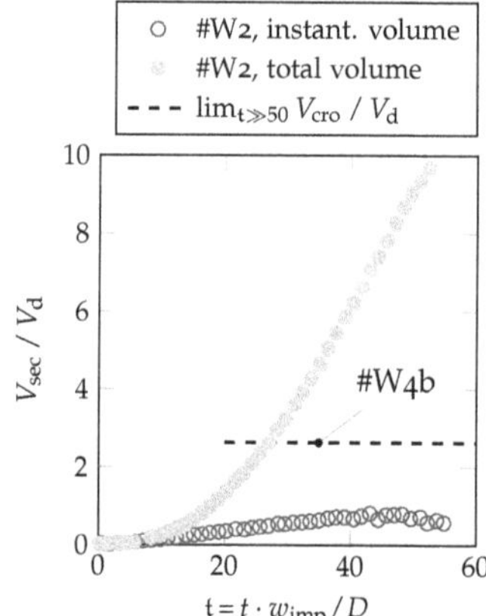

Figure 7.10: Increasing instantaneous volume of secondary droplets over time inside the FOV of the shadowgraph camera. After mass-flux weighted integration over time, the total volume of the secondary droplets is obtained. The volume of the entire crown is plotted for comparison.

According to the breakup process described in the previous section, the entire crown is atomized into secondary droplets. V_{sec} and V_{cro} shall thus be on the same order of magnitude. Figure 7.10 compares the two results. Notably, the total volume of the secondary droplets is at least twice the primary drop volume. Wet impact creates significantly more ejected mass than dry impact because the lamella consists of both drop and film fluid. FASSMANN (2015) predicts even more secondary volume. Since $V_{sec} \gg V_{cro}$, one might argue that a non-negligible uncertainty arises in his mass flux estimations.

Part III

Fluid mechanics of icing at large scales

Overview

Macroscopic ice accretion shall be studied by a combined approach of computations and experiments. For this purpose, a low-cost icing wind tunnel was built in Braunschweig that promotes further advancement of both fundamental and applied icing research. At the same time, a computational tool to predict ice accretion was developed in collaboration with the German Aerospace Centre (DLR). Joining these capabilities will yield a deeper understanding of the complex process of glaze ice accretion, which involves many of the small-scale physical phenomena mentioned in the previous part of this manuscript.

8
Icing wind tunnel

Icing wind tunnels are a convenient tool to reproduce the physics of aircraft icing in a controlled laboratory environment. To promote further advancement in both fundamental and applied icing research, a low-cost icing wind tunnel was built in Braunschweig, and the design considerations are presented in this chapter. Furthermore, selected experimental techniques for characterizing the droplet cloud and the ice accretion are introduced.

8.1 *Are icing wind tunnels still needed?*

"In the 1970s and continuing into the early 1980s a sizable group of experts predicted that the need for aerodynamic experiments, particularly in the subsonic regime, would rapidly disappear as computational fluid dynamics would in rather short time become sufficiently capable so that all needed information would be available from computational simulations at a cost-effectiveness superior to that of experiments." — a statement that can be found in the preface of "Low-speed wind tunnel testing" by Barlow *et al.* (1999).

Today, nearly half a century later, the computational capacity has increased by a factor of approximately 10^9.[1] Simulations of the Reynolds-averaged Navier-Stokes equations are now standard in the aeronautical community. Eddy-resolving simulations have found their way into industrial design offices, to some extent even being predictive. Do we fully trust these simulations in significant development projects? Turbulence still challenges us in many respects.

[1] A Cray-1, first installed in 1976 at Los Alamos National Laboratory, performed at 160 MFLOPS using 115 kW. In 2016, the fastest supercomputer is the 神威·太湖之光, with 93 PFLOPS and 15 MW.

Aircraft icing is subject to the same fate. The droplet impact, water film dynamics and heat transfer along surfaces with anisotropic roughness significantly increase the complexity of this multi-scale problem. It is therefore not surprising that icing wind tunnel testing remains an essential part in technology maturation and aircraft certification. Ensuring flight safety is one of the leading priorities in this context. Figure 8.1 underlines a general trend: the prevalence

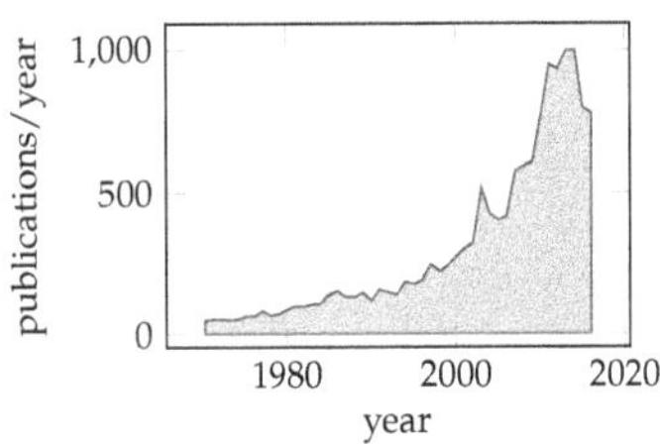

Figure 8.1: Result of a full-text query for the search pattern *icing wind tunnel* on Google Scholar.

of the topic *icing wind tunnel* in the scientific literature has grown exponentially over the past few decades.

8.2 *Building the Braunschweig Icing Wind Tunnel*

Overall design

A schematic drawing of the Braunschweig Icing Wind Tunnel[2,3] is shown in Figure 8.2. The overall objective in icing wind tunnel design is to obtain a homogeneous distribution of the flow velocity, temperature and icing cloud throughout the test section ①. The requirement of uniformity in velocity generally results in an apparatus that is similar to conventional closed-loop wind tunnels. Features such as the diffuser ②, turning vanes ⑤, settling chamber ⑦ with screens and honeycombs, contraction ⑨ and fan ③ are familiar. The design guidelines for these classical elements can be found in Barlow *et al.* (1999).[4]

[2] Stephan E. Bansmer, Arne Baumert, Stephan Sattler, Inken Knop, Delphine Leroy, Alfons Schwarzenboeck, Tina Jurkat-Witschas, Christiane Voigt, Hugo Pervier, and Biagio Esposito. Design, Construction and Commissioning of the Braunschweig Icing Wind Tunnel. *Atmospheric Measurement Techniques*, 11: 3221–3249, 2018

[3] This section is a short summary of the above publication and might reuse some of its phrases.

[4] Jewel B. Barlow, William H. Rae, and Alan Pope. *Low-speed wind tunnel testing*. Wiley, New York, NY [u.a.], 3. ed edition, 1999. ISBN 0471557749

In this context, the determination of the fan power is noteworthy. Given an air density of ϱ_{air}, a maximum speed U_∞, and a cross-sectional area A_{ts} of the tunnel test-section, the required jet power P_{jet} can be calculated as

$$P_{\text{jet}} = \frac{1}{2}\varrho_{\text{air}} U_\infty^2 \cdot (U_\infty \cdot A_{\text{ts}}) = \frac{1}{2}\varrho_{\text{air}} U_\infty^3 \cdot A_{\text{ts}} \quad .$$

To obtain the fan power P_{fan}, the jet power P_{jet} is multiplied by a power factor $\mathfrak{P}$. Conventional closed-loop, low-speed wind tunnels have power factors of approximately $\mathfrak{P} \approx 1.5$. The fan power has to compensate for the pressure and skin friction losses of the individual tunnel components. High losses are caused by the wall boundary layers of the test section and the tunnel diffuser. The first and second corners after the diffuser also create significant losses. In icing wind tunnels, the vanes of the first corner are subject to ice accretion. Their losses are much higher than those of conventional closed-loop tun-

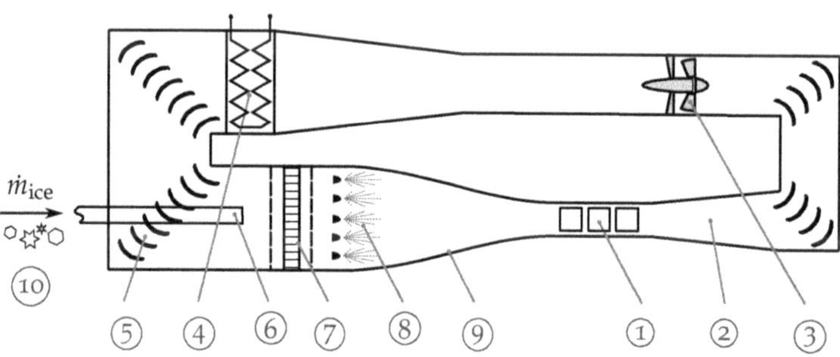

Figure 8.2: Schematic overview of the Braunschweig Icing Wind Tunnel. Beyond the icing phenomena of supercooled droplets, ice crystal icing can also be studied. For this purpose, replicas of natural ice crystals (10) are injected into the tunnel circuit using a pneumatic conveyance (6). The screens and honeycombs (7) are removed for this type of test. See Baumert et al. (2018) for further information.

nels. Consequently, the power factor of the Braunschweig Icing Wind Tunnel was dimensioned to a large value of $\mathfrak{P} = 2.3$.

Spray system and droplet trajectories

What makes a wind tunnel an icing wind tunnel? Predominantly, it is the spray system (Figure 8.2-⑧), which generates small water droplets that are seeded into the air flow. Typically, pneumatic atomizers are used for this task. In varying the applied water and gas pressures at these atomizers, a broad spectrum of spray conditions can be achieved, varying in both size and concentration. An informative introduction to the field of liquid atomization can be found in BAYVEL (1993).[5]

[5] Leoplod P. Bayvel and Zdzisław Orzechowski. *Liquid atomization*. Combustion. Taylor & Francis, Washington, 1993. ISBN 0891169598

HAVING LEFT the atomizer, the droplets enter the tunnel contraction (Figure 8.2-⑨), where the air flow is accelerated. Two different contraction geometries were considered during the design phase. These are the AVA nozzle, with its contour given by

$$\frac{z(x)}{z_{\text{ts}}} = \left(\sqrt{\mathfrak{C}} - 1\right)\left(1 - \frac{x}{l}\right)^3 \left[2 - \left(1 - \frac{x}{l}\right)^3\right] + 1 \text{, and}$$

the WITOSZYNSKI nozzle, with its contour given by

$$\frac{z(x)}{z_{\text{ts}}} = \frac{1}{\sqrt{1 + \left(\frac{1}{\mathfrak{C}} - 1\right) \frac{\left[1 - \left(\frac{x}{l}\right)^2\right]^2}{\left[1 + \frac{1}{3}\left(\frac{x}{l}\right)^2\right]^3}}} \quad ,$$

where x, z are spatial coordinates, l is the nozzle length, $z_{\text{in}}, z_{\text{ts}}$ are half of the inlet and outlet diameters of the nozzle, and the contraction $\mathfrak{C}$ is given by $\mathfrak{C} = \left(\frac{z_{\text{in}}}{z_{\text{ts}}}\right)^2$. To provide enough time for the super-cooling process of the droplets on their way to the tunnel test section, a large nozzle length of 3.5 m is employed.

THE DROPLET TRAJECTORIES can be computed using a balance of the inertial forces, drag and gravity. A droplet with diameter d, velocity $\vec{u}_{\text{d}} = \binom{u_{\text{d}}}{w_{\text{d}}}$ is conveyed by the air flow with velocity $\vec{u}_{\text{air}} = \binom{u_{\text{air}}}{w_{\text{air}}}$. The drag force is

$$\vec{F}_{\text{D}} = \frac{\varrho_{\text{air}}}{2} \left|\Delta\vec{u}\right|^2 d^2 c_D \cdot \frac{\Delta\vec{u}}{\left|\Delta\vec{u}\right|} \quad ,$$

where $\Delta\vec{u} = \vec{u}_{\text{air}} - \vec{u}_{\text{d}}$ denotes the slip velocity. Assuming laminar flow around the spherical droplets at low Reynolds numbers $\text{Re}_{\text{air}} = \frac{\varrho_{\text{air}} \cdot |\Delta\vec{u}| \cdot d}{\mu_{\text{air}}}$, the drag coefficient c_D can be approximated as $c_D = \frac{6\pi}{\text{Re}_{\text{air}}}$. Application of the principle of d'Alembert yields a system of differential equations, given by

$$m_{\text{d}} \frac{\partial \vec{u}_{\text{d}}}{\partial t} = \vec{F}_{\text{D}} + m_{\text{d}}\vec{g} \quad ,$$

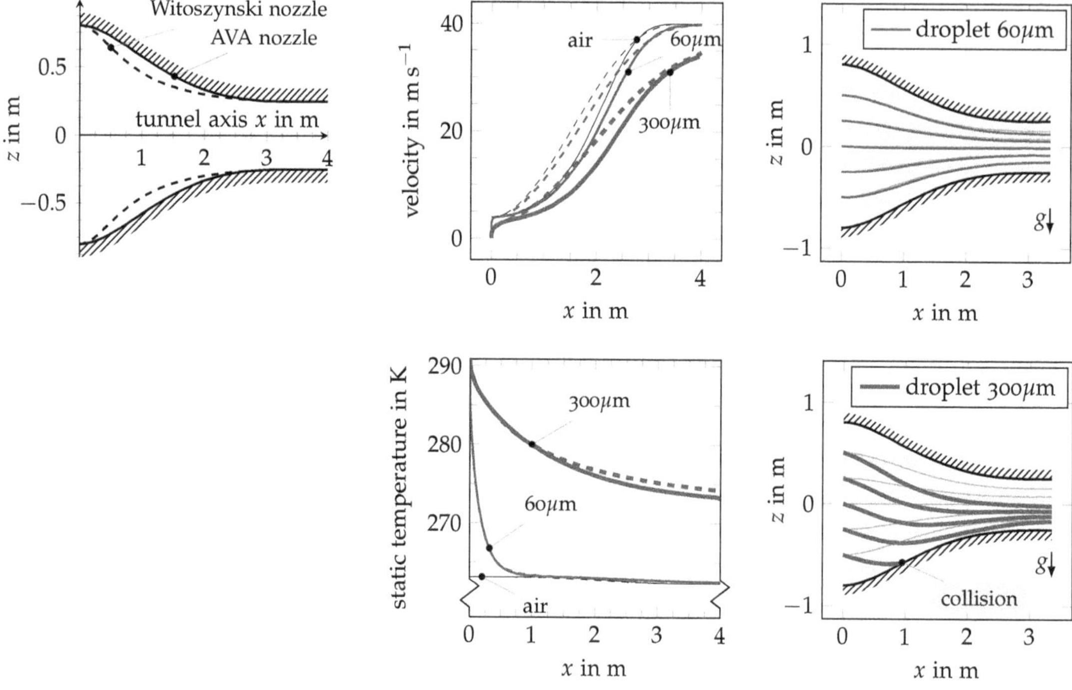

Figure 8.3: Velocity and temperature slopes for air and two different-sized droplets inside two tunnel nozzle configurations *(left & middle)*. Contraction of the droplet trajectories in the wind tunnel and their deviation from the air streamlines under the influence of gravity. Large droplets may collide with the tunnel walls, thereby altering the droplet size distribution *(right)*.

where t represents the time, m_d denotes the droplet's mass and $\vec{g}$ is the acceleration due to gravity.

THE EFFECT OF GRAVITY on the droplet trajectories is illustrated in Figure 8.3, assuming an air velocity of $40\,\mathrm{m\,s^{-1}}$ at the nozzle exit. Small droplets with a size of 60 µm follow the air flow streamlines of inside the nozzle. With increasing droplet size, the gravitational deflection predominates, particularly in the low-speed regions of the nozzle. For 300 µm droplets, their trajectory from the lowest spray-bar may even collide with the wall of the contraction. Such undesired collisions alter the size distribution of the droplet cloud and should be avoided.

FIGURE 8.3 further demonstrates the effect of the nozzle geometry on the droplet slip velocity. Note that after the nozzle length of $l = 3.5\,\mathrm{m}$, a straight segment of 0.5 m in length is attached. In the low-speed region $x < 1.5\,\mathrm{m}$, the AVA nozzle accelerates the flow less than the Witoszynski nozzle. Consequently, the slip in velocity between the droplet and air is larger in that region. Downstream, for $x > 3\,\mathrm{m}$, the slip is caught up. While the small droplet of 60 µm approaches an air speed of $40\,\mathrm{m\,s^{-1}}$ when entering the test section at $x = 4\,\mathrm{m}$, the large droplet of 300 µm still has a slip of approximately $5\,\mathrm{m\,s^{-1}}$.

FURTHERMORE, THE COOLING PROCESS of the droplet on its trajectory is considered. Cooling is promoted by convective heat transfer at the interface between the droplet and air. When the droplet is smaller than 100 µm, the surface temperature penetrates the entire droplet volume in approximately 10 milliseconds. Hence, isothermal cooling can be assumed as an approximation for the thermal energy balance, yielding a differential equation for the droplet temperature T_d:

$$\frac{\pi}{6} d^3 \varrho_d c_{\mathrm{p,\,d}} \frac{\partial T_d}{\partial t} = \mathrm{Nu} \cdot \pi d^2 \left(T_{\mathrm{air}} - T_d\right) \cdot \frac{k_{\mathrm{air}}}{d} \quad ,$$

where ϱ_d and $c_{\mathrm{p,\,d}}$ represent the density and the specific heat capacity of the droplet, respectively, k_{air} is the heat conductivity of air, and Nu is the Nusselt number, estimated by $\mathrm{Nu} = 2 + 0.6\,\mathrm{Pr}^{\frac{1}{3}}\,\mathrm{Re}_{\mathrm{air}}^{\frac{1}{2}}$.[6] Herein, $\mathrm{Pr} = 0.7$ is the Prandtl number of air. Using the above equation, the temperature evolution of the droplet inside the tunnel nozzle is plotted in Figure 8.3 for a given static air temperature of $T_{\mathrm{air}} = 263.15\,\mathrm{K}$. After a one-metre travelling distance, the droplet of 60 µm, which had an initial temperature of $T_{\mathrm{air}} = 293.15\,\mathrm{K}$, has already reached air temperature. In contrast, the droplet of 300 µm still has a temperature above the freezing point of water when entering the test section at $x = 4\,\mathrm{m}$.

[6] James G. Knudsen and Donald L. Katz. *Fluid dynamics and heat transfer.* McGraw-Hill chemical engineering series. McGraw-Hill, New York [u.a.], 1958

In summary, droplet conditions up to a maximum size of 150 µm can be accurately provided in the present tunnel environment, taking into account a reasonably low deflection due to gravity, a low slip velocity, and sufficient droplet cooling.

Chilling power exceeds fan power

There are three major heat sources that have to be compensated for by the heat exchanger (Figure 8.2-④) and chilling device: the power of the tunnel fan, heat input through the wind-tunnel walls, and heat transfer of the water spray. The last is composed of the *sensible heat* $\dot{Q}_{\mathrm{sens}}$ to super-cool the water spray from the temperature at which it leaves the pneumatic atomizer (approximately 20 °C) down to the air temperature inside the tunnel test-section and the *latent heat* $\dot{Q}_{\mathrm{fus}}$. As soon as the supercooled water droplet impacts on a solid substrate, the drop solidifies due to heterogeneous nucleation and releases latent heat. Both the sensible and latent heats are functions of the liquid water content (LWC)

$$\begin{aligned}\dot{m}_{\mathrm{w}} &= \mathrm{LWC} \cdot \dot{V}_{\mathrm{air}} \\ &= \mathrm{LWC} \cdot (U_\infty \cdot A_{\mathrm{ts}}) \\ \dot{Q}_{\mathrm{sens}} &= \dot{m}_{\mathrm{w}} \cdot c_{\mathrm{p,w}} \cdot \Delta T \\ \dot{Q}_{\mathrm{fus}} &= \dot{m}_{\mathrm{w}} \cdot L_{\mathrm{fus}} \quad ,\end{aligned}$$

where $\dot{m}_{\mathrm{w}}$ is the water mass flow rate of the spray atomizers and ΔT the temperature difference, with the following values for demineralized water:

$$\begin{aligned}c_{\mathrm{p,w}} &= 4.183 \frac{kJ}{kg \cdot K} \\ L_{\mathrm{fus}} &= 334 \frac{kJ}{kg} \quad .\end{aligned}$$

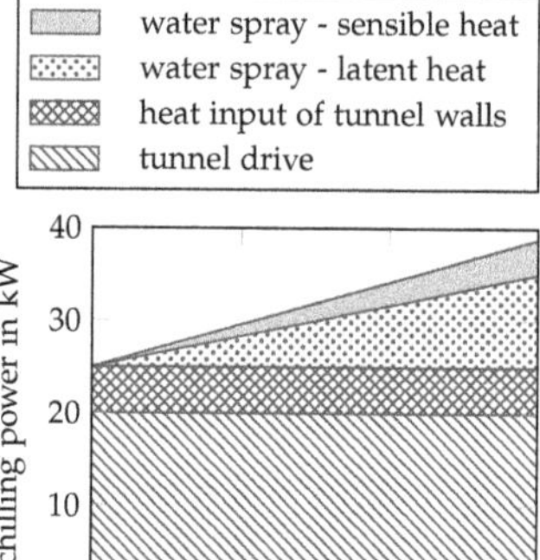

Figure 8.4: Required chilling power decomposition as a function of the LWC for typical operational conditions of the Braunschweig Icing Wind Tunnel.

For the typical operational conditions of the icing wind tunnel, i.e., $U_\infty = 40\,\mathrm{m\,s^{-1}}$ and $\Delta T = 30\,\mathrm{K}$, and approximations for the fan power and the heat input through the wind-tunnel walls, the necessary cooling capacity can be estimated; see Figure 8.4. When LWC is in the range of $1\,\mathrm{g\,m^{-3}}$, a continuous cooling power of 30 kW is necessary. To allow for a five-minute peak load of LWC $= 3\,\mathrm{g\,m^{-3}}$ and static tunnel temperatures down to $-20\,°\mathrm{C}$, the system was dimensioned for a maximum cooling power of 80 kW.

8.3 *Calibration of the Braunschweig Icing Wind Tunnel*

The main design targets of an icing wind tunnel are to provide both homogeneous airflow and a uniform droplet cloud inside the test section. The calibration step aims to compare these targets with the actual conditions.[7]

[7] AC-9C Aircraft Icing Technology Committee. Calibration and Acceptance of Icing Wind Tunnels. Technical report, SAE International, 2015

Aerothermal calibration

An aerothermal calibration is conducted to demonstrate adequate airflow quality. The airflow uniformity is investigated by a Prandtl probe and shows only deviations ΔU smaller than $\pm 1\%$ from the mean velocity U_∞ in the centre of the test section; see Figure 8.5. The turbulence intensity is measured using hot-wire anemometry.[8] Both the aerodynamic wake of the spray system and the gas jets of the pneumatic atomizers create a significant turbulence level between 1% and 2%, which is deemed necessary to provide a satisfactory dispersion of the droplets in the flow. The temperature distribution is determined using a total temperature probe. The spatial variation is smaller than $\pm 0.5\mathrm{K}$.

[8] Genevieve Comte-Bellot. Hot-wire anemometry. *Annual review of fluid mechanics*, 8(1):209–231, 1976

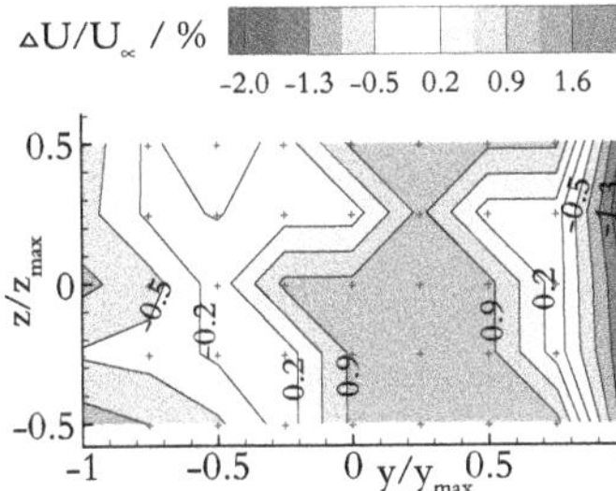

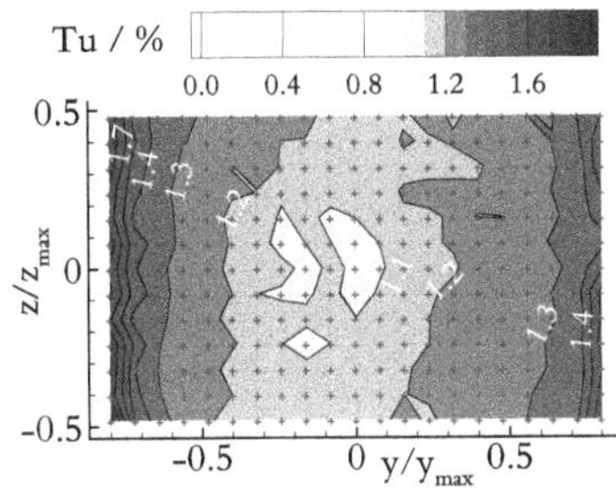

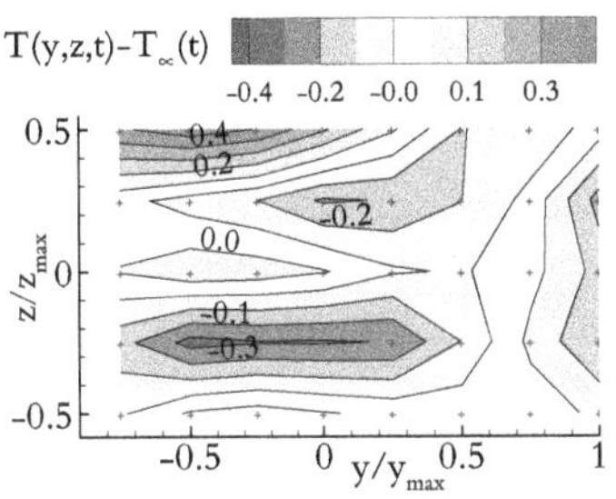

Figure 8.5: Aerothermal calibration data across the test section of the Braunschweig Icing Wind Tunnel for $U_\infty = 40\text{m/s}$ and $T_\infty = -10°\text{C}$. Data from INKEN KNOP.

Optical characterization of droplets

Optical measurement techniques offer an elegant approach to characterize the droplet cloud inside the icing wind tunnel in terms of size and concentration. The mathematical description of the underlying background is based on the work of LORENZ and MIE, which applies MAXWELL's electromagnetic theory to describe the scattering of light by small homogeneous spheres.[9] Geometrical optics can provide a useful approximation when the particle is sufficiently large relative to the wavelength of the incident light. See Figure 8.6 for a two-dimensional example. When the light ray first hits the surface, both a reflected and a refracted ray emanate with intensities that follow the Fresnel coefficients. When the refracted ray again reaches the surface, a similar separation between reflection and refraction occurs, which is called *first-order refraction*. This process continues ad infinitum. For a three-dimensional sphere, the points where the refracted rays leave the particle are called glare points. Note that only half of the total scatted intensity arises from reflection and refraction; the other half is due to diffraction, which is one of the most challenging optical phenomena to model and cannot be described by the laws of geometrical optics.

[9] Hendrik C. van de Hulst. Light Scattering by Small Particles, 2012

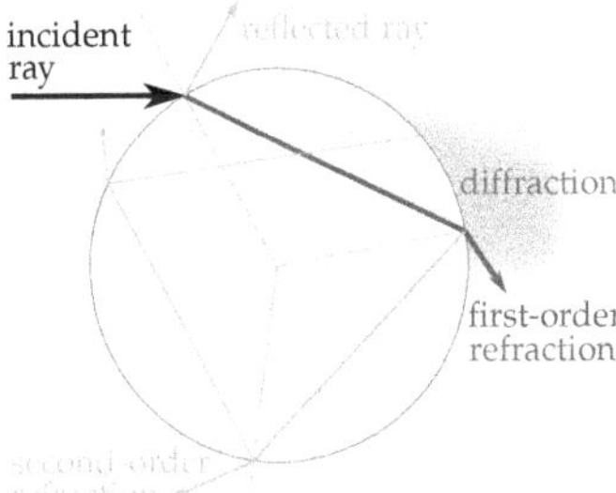

Figure 8.6: Light rays through a circle according to geometrical optics.

MANY MEASUREMENT TECHNIQUES for particle characterization can be developed with this background in mind.[10] The *direct imaging* of particles is probably the most intuitive technique; shadowgraphy, as a prominent example, was already used for the droplet impact studies presented in Part II of this book. *Interferometry* represents a second category of measurement techniques, of which phase Doppler interferometry (PDI)[11] is the most established one for particle sizing. In its standard arrangement, two intersecting laser beams with wavelength λ_b are used; see Figure 8.7. When a particle traverses the detection volume in the x-direction, it exhibits two individual glare points of first-order refraction. We find that the glare point distance is inter alia a function of the particle diameter d_P. Moreover, the glare

[10] Cameron Tropea. Optical particle characterization in flows. *Annual Review of Fluid Mechanics*, 43:399–426, 2011

[11] Heinz-Eberhard Albrecht. Laser Doppler and Phase Doppler Measurement Techniques, 2003

points can be interpreted as two coherent point light sources that produce interference patterns in the far field. Two photodetectors R_1 and R_2 detect the interference fringes but are pivoted with an angle Φ from the forward scattering x-z-plane, thus avoiding the detection of diffracted light. Due to their symmetric elevation Ψ from the y-z-plane, the photodetectors receive the interference fringes with a phase shift of

$$\Delta\varphi_{\mathrm{PDI}} \approx -\frac{4\pi}{\lambda_b} d_P \frac{\mathfrak{n} \sin\Psi \sin\frac{\Theta}{2}}{C\sqrt{1+\mathfrak{n}^2 - \mathfrak{n}C}}$$

$$\text{with} \quad \mathfrak{n} = \frac{n_P}{n_{\mathrm{gas}}} \quad C = \sqrt{2\left(1+\cos\Psi \cos\Phi \cos\frac{\Theta}{2}\right)} \quad ,$$

where n_P and n_{gas} represent the refractive indexes of the particle and surrounding gas, respectively. In essence, the phase shift and particle diameter are related by the geometrical quantities of the configuration, which significantly minimizes the calibration procedure.

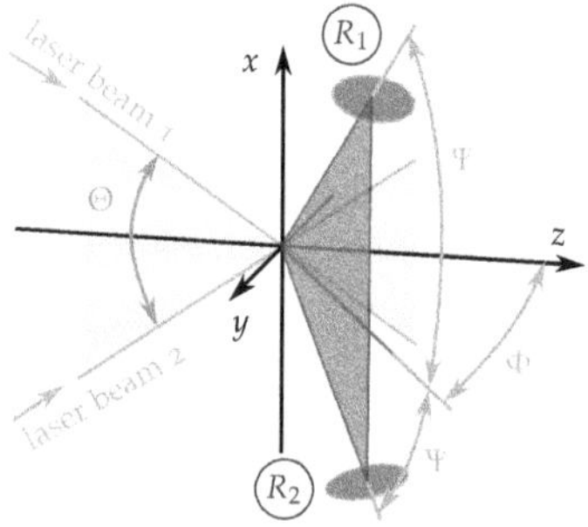

Figure 8.7: Standard configuration for PDI. Two receivers R_1 and R_2 record laser light scattered by particles traversing the detection volume.

Droplet cloud calibration results

Beyond PDI, additional techniques are also used to quantify the droplet size distribution and the LWC; these include the time-shift technique,[12] the fast cloud droplet probe (FCDP)[13], shadowgraphy using a similar configuration to those described in part II, and a rotating cylinder.[14]

THE DROPLET CLOUD ENVELOPE of the Braunschweig Icing Wind Tunnel can be deduced from the calibration; see Figure 8.8 *(lower left)*. The size distribution of the droplet cloud is reduced to the scalar MVD, and the droplet concentration is represented by LWC. The tunnel's spray system can generate cloud conditions in the MVD range between 10 and 50 µm and in the LWC range between 0.1 and 2 gm^{-3}. This envelope lies well within the certification requirements of FAR25/CS25 Appendix C.

MOREOVER, AN INTER-COMPARISON between the different measurement techniques is possible since many test points have been repeatedly investigated. Figure 8.8 *(upper left)* shows the inter-comparison for MVD using the PDI measurements as a reference. A very high correlation appears between the results for the whole range of the envelope; a measurement uncertainty of ±10 µm is anticipated. More specifically, shadowgraphy provides satisfactory agreement for a high MVD since the direct imaging technique benefits from the high optical resolution of individual large droplets. On the other hand, the FCDP is known to provide accurate sizing information

[12] Walter Schäfer and Cameron Tropea. Time-shift technique for simultaneous measurement of size, velocity, and relative refractive index of transparent droplets or particles in a flow. *Applied Optics*, 53(4):588–597, 2014

[13] www.specinc.com

[14] James R. Stallabrass. An appraisal of the single rotating cylinder method of liquid water content measurement. Technical report, National Research Council of Canada (NRC), 1978

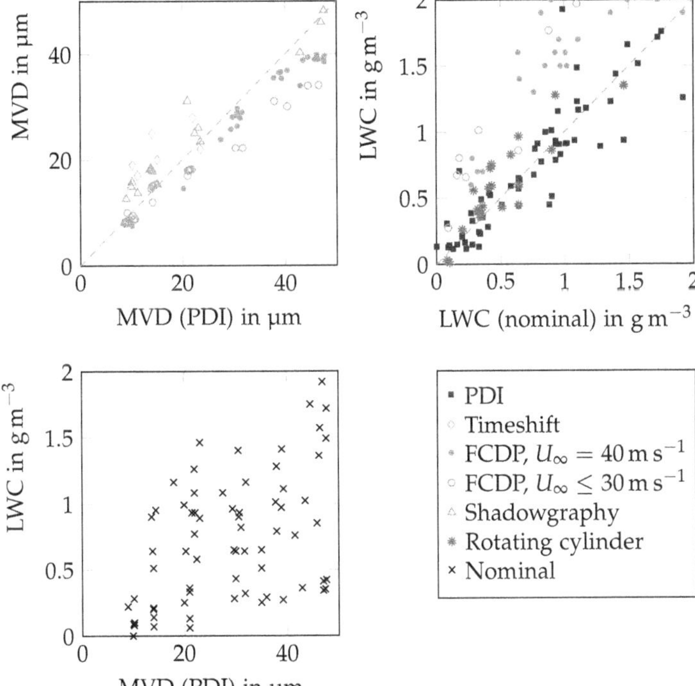

Figure 8.8: Droplet cloud calibration results and probe inter-comparison. *Upper left:* MVD comparison of various measurement techniques against PDI results. *Upper right:* LWC comparison of various measurement techniques against its estimation from the water flow rate through the atomizers. *Lower left:* MVD-LWC envelope of the Braunschweig Icing Wind Tunnel at $U_\infty = 40\,\mathrm{m\,s^{-1}}$. Data courtesy of Inken Knop, Valerian Hahn (DLR), Jonas Kleine (DLR), AOM Systems and LaVision.

for small droplets in the diameter range from 1 µm to 50 µm, i.e., MVD $\leq$ 20 µm. Notably, a velocity dependence appears within the FCDP results. The time-shift measurements are still preliminary because both velocity and LWC dependencies were recently discovered in the MVD; these aspects are under current investigation.

The uncertainties in the LWC measurements are presented in Figure 8.8 *(upper right)*. The mass flow rate through the spray system $\dot{m}_\mathrm{W}$ and the effective area A_eff covered by the droplet cloud in the test section are used to estimate the nominal LWC as follows:

$$\mathrm{LWC}_\mathrm{nominal} = \frac{\dot{m}_\mathrm{W}}{U_\infty \cdot A_\mathrm{eff}} \quad ,$$

against which the LWC results from different techniques are plotted. Generally, a correlation is visible, but the uncertainties are significantly larger than those of the MVD measurements. More than 100 % uncertainty is sometimes evident, which was previously reported for optical mass-flux measurements.[15] One of the major sources of error is the a priori unknown reference volume that is needed for computing the mass flux of the droplets – and thus the LWC. Indeed, the reference volume may even be a function of the droplet

[15] Cameron Tropea. Optical particle characterization in flows. *Annual Review of Fluid Mechanics*, 43:399–426, 2011

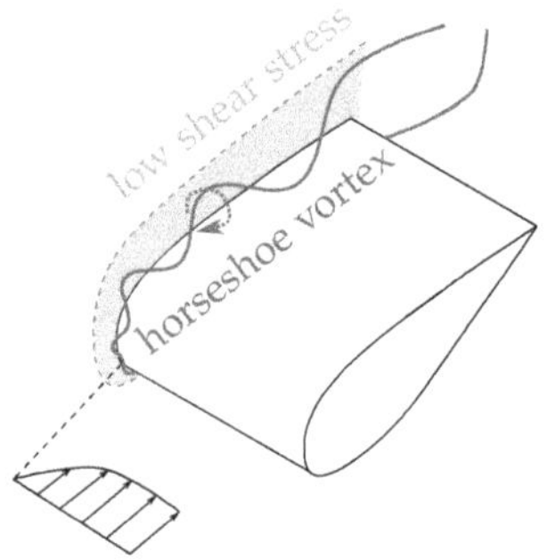

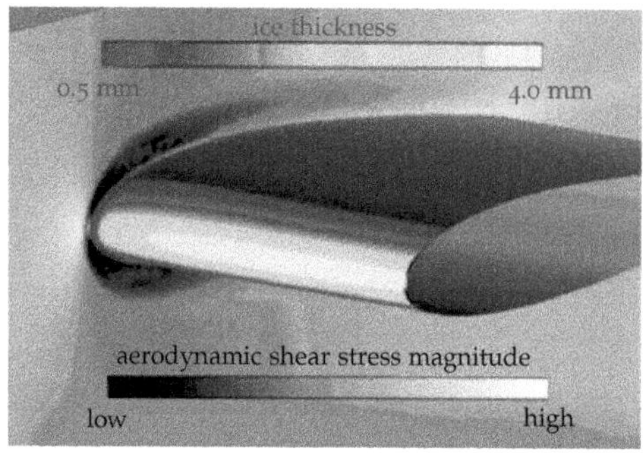

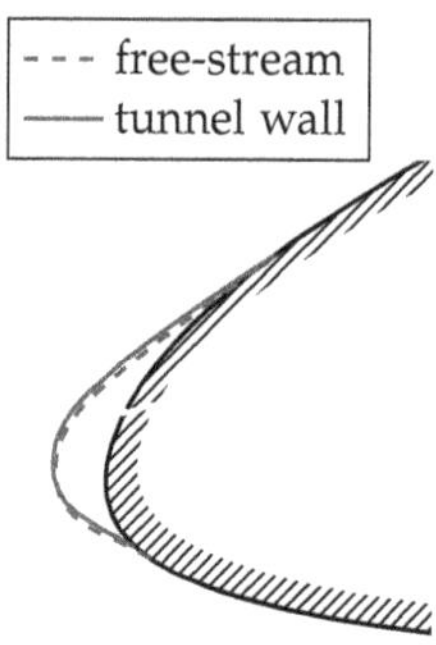

Figure 8.9: Wind tunnel wall effects on ice accretion. *Left:* Horseshoe vortex system. Sketch inspired by SIMPSON (2001). *Centre:* The junction flow between a NACA0012 aerofoil ($c = 0.75\,\mathrm{m}$, AoA $= 0°$) and the wind-tunnel side wall alters ice accretion. $U_\infty = 40\,\mathrm{m\,s^{-1}}$, $T_\infty = -10\,°\mathrm{C}$, LWC=$1\,\mathrm{g\,m^{-3}}$. *Right:* Inviscid wall effect on ice accretion for an AH-94-W-145 aerofoil, $c = 0.75\,\mathrm{m}$, U_∞=$40\,\mathrm{m\,s^{-1}}$, $T_\infty = -10\,°\mathrm{C}$, LWC=$0.3\,\mathrm{g\,m^{-3}}$, MVD=$15\,\mu\mathrm{m}$. Data from BANSMER et al. (2018).

diameter itself, hence requiring the careful calibration of the optical systems. A further hypothesis involving a dependence between the reference volume particle velocity is currently being investigated, which underlines that optical measurement techniques for particle characterization are still an active field of research.

8.4 *Wind tunnel wall effects on ice accretion*

Minimizing the experimental uncertainties while applying scaling laws for ice accretion typically requires wind tunnel models with large dimensions. Thereby, the distance to the adjacent wind tunnel walls decreases, and the aerodynamics hence deviate from free-stream conditions. Two important wall effects are prevalent.

JUNCTION FLOW[16] is the first effect; see Figure 8.9 (left). The boundary layer of the wind tunnel wall encounters the leading edge of the test model attached to the same wall. An adverse pressure gradient thus arises that forces the wall boundary layer to separate. The main flow wraps the emerging vortex around the obstacle and creates a horseshoe vortex. Its unsteadiness causes high turbulence intensities, alters the trajectories of the droplets and further affects the heat transfer rates. The influence of the junction flow on the ice accretion was computationally studied using the icing code FENSAP;[17] see Figure 8.9 (centre). The footprint of the horseshoe vortex can be clearly identified by the wall shear stress distribution. A glaze ice case was simulated with boundary conditions of temperature $T_\infty = -10\,°\mathrm{C}$, LWC=$1\,\mathrm{g\,m^{-3}}$, and ice accumulation time $t_{acc} = 120\,\mathrm{s}$. The velocity field influences both the local collection efficiency of the water droplets and the heat transfer. Consequently, the ice accretion is attenuated at the junction. Nevertheless, the two-dimensional ice shape farther from the tunnel wall, covering approximately 80% of the wing span, remains unaffected.

[16] Roger L. Simpson. Junction flows. *Annual Review of Fluid Mechanics*, 33 (1):415–443, 2001. DOI: 10.1146/annurev.fluid.33.1.415

[17] Héloïse Beaugendre, François Morency, and Wagdi G. Habashi. FENSAP-ICE's Three-Dimensional In-Flight Ice Accretion Module: ICE3D. *Journal of Aircraft*, 40(2):239–247, March 2003. DOI: 10.2514/2.3113

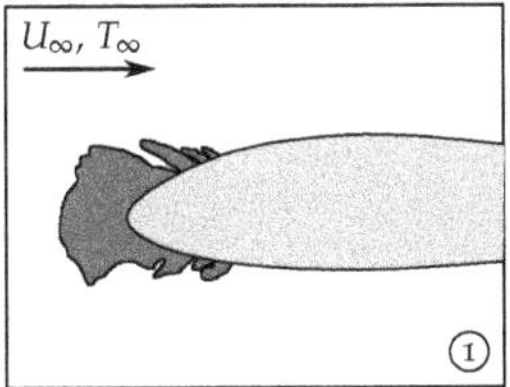

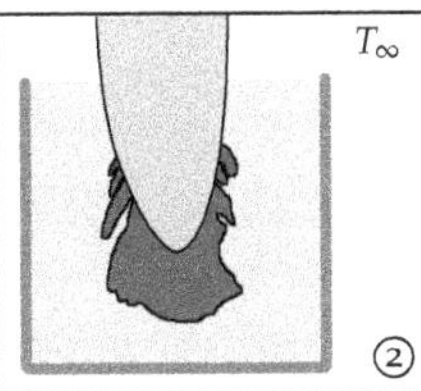

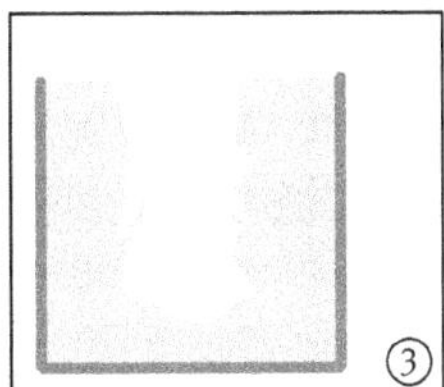

Figure 8.10: Processing steps to create a polyurethane replica of accreted ice.

THE DISTANT WALLS that are not connected to the test model cause an inviscid wall effect by changing the curvature of the airflow streamlines. Therefore, the distributions of the pressure and shear stress along the test model's surface are modified. Computations with TAUICE – an icing code presented in Chapter 9 – reveal that for low airspeeds U_∞ and adequately large droplet sizes (MVD=15 µm), the ice accretion remains almost unaffected when tunnel walls are introduced; see Figure 8.9 (right).

8.5 *Measuring the geometry of ice accretion*

The classical approach: mould and cast

Recording the filigree shapes of accreted ice is a challenging task: the roughness elements are extremely fragile, and temperature changes lead to deterioration. A method for the long-term preservation of ice shapes that avoids handling the ice itself can be accomplished by the mould-and-cast method; see Figure 8.10. The general approach has long been known;[18] new materials[19] facilitate the process and enable faithful reproduction.

[18] Andrew L. Reehorst and G. Paul Richter. New Methods and Materials for Molding and Casting Ice Formations. *NASA Technical Memorandum*, (100126), 1987

[19] Mold Star 16 and Smooth-Cast 300 are currently the predominant commercial products.

FIRST, A NEGATIVE MOULD of the accreted ice is created ② using a low-temperature, addition-cure silicone with low shrinkage. For this purpose, the silicone fluid is mixed with a proper corresponding catalyst and then poured into a mould box. The mixture is stored for one hour at cold temperatures, allowing large air bubbles that were entrapped during mixing to float to the surface. Thereafter, the iced aerofoil is carefully inserted. While curing for approximately 18 hours, it is important to maintain the temperature at which the ice was initially accreted. After demoulding and drying ②, a polyurethane resin is used to cast a positive replica of the ice shape ④. The last demoulding step ⑤ is typically impeded by the undercutting of ice feathers.

Optical approach: penetration of light into water and ice

To avoid any mechanical treatment for the determination of ice shapes, optical measurement techniques seem promising. Before a specific technique is presented, the interaction of light with water and ice is briefly introduced.

WATER HAS A SMALL, yet non-negligible, electrical conductivity γ_{el} on the order of 10^{-6}—10^{0} S/m depending on its purity level. Light as an electromagnetic wave interacts with conducting media. Its energy is transferred into heat, which attenuates the light. This effect can be mathematically described by introducing a complex index of optical refraction $\hat{n} = n_{real} + i \cdot n_{imag}$ instead of a real one. Following the textbook of MAX BORN,[20] Maxwell's equations are transformed into a wave equation

[20] Max Born and Emil Wolf. *Principles of optics*. Cambridge University Press, 7., (expanded) ed., reprinted with corr. edition, 2002. ISBN 9780521642224

$$\nabla^2 \vec{E} + \underbrace{\frac{\mu_{mag}\omega^2}{c_0^2}\left(\varepsilon_{el} + i\frac{4\pi\gamma_{el}}{\omega}\right)}_{=\hat{k}^2} \vec{E} = \vec{0} \quad ,$$

Given the electric vector $\vec{E}$, the magnetic permeability μ_{mag}, the dielectric constant ε_{el}, the speed of light in vacuum c_0, the angular frequency ω and the imaginary unit i, the complex wave number $\hat{k}$ is related to the complex index of refraction $\hat{n} = \frac{c_0}{\omega}\hat{k}$. For a monochromatic plane wave

$$\vec{E} = \vec{E}_0 e^{i\left[\hat{k}x - \omega t\right]} = \vec{E}_0 e^{i\left[\frac{\omega}{c_0}\left(n_{real} + i \cdot n_{imag}\right)x - \omega t\right]} = \vec{E}_0 e^{i\left[\frac{\omega n_{real}}{c_0}x - \omega t\right]} e^{-\frac{\omega n_{imag}}{c_0}x} ,$$

the energy density $w_{\vec{E}}$ is proportional to the time average of $\vec{E}^2$, yielding

$$w_{\vec{E}} \sim \vec{E} \cdot \vec{E}^{\star} \sim \vec{E}_0^2 e^{-\frac{2\omega n_{imag}}{c_0}x} \quad .$$

The exponent $\frac{2\omega n_{imag}}{c_0}$ is called the absorption coefficient; its reciprocal represents the distance d that light has travelled until the time that its energy density has fallen to $1/e$ of its initial value. d is plotted over the wavelength of light for water[21] and ice Ih[22] in Figure 8.11. Visible light can deeply penetrate the (clear) ice, while infrared radiation is absorbed within a few micrometres, which makes it advantageous for a laser scanning technique.

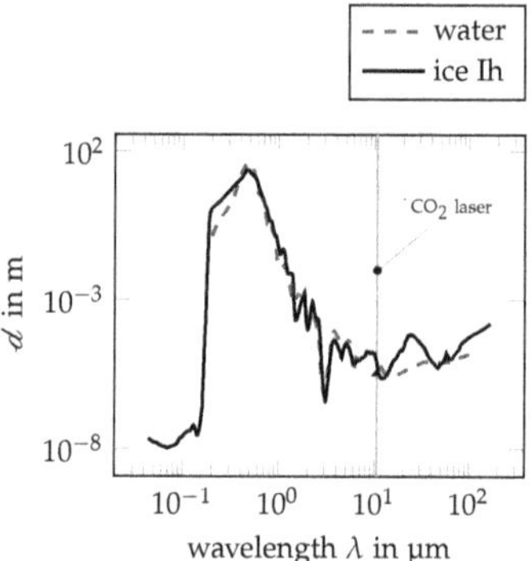

Figure 8.11: Penetration depth of light of different wavelengths λ into water and ice Ih.

[21] George M. Hale and Marvin R. Querry. Optical constants of water in the 200-nm to 200-μm wavelength region. *Applied optics*, 12(3):555–563, 1973

[22] Stephen G. Warren. Optical constants of ice from the ultraviolet to the microwave. *Applied Optics*, 23(8): 1206–1225, 1984

Infrared laser line scanning of ice

A variety of optical techniques can be used to capture ice surface geometries.[23] Because of its simplicity and robustness, the line scanning method is the most widely used triangulation-based scanning technique for 3D geometries.[24] Figure 8.12 (left) shows the general

[23] Xiaoliang Gong and Stephan E. Bansmer. Laser scanning applied for ice shape measurements. *Cold Regions Science and Technology*, 115:64–76, 2015b

[24] François Blais. Review of 20 years of range sensor development. *Journal of Electronic Imaging*, 13(1):231–243, 2004

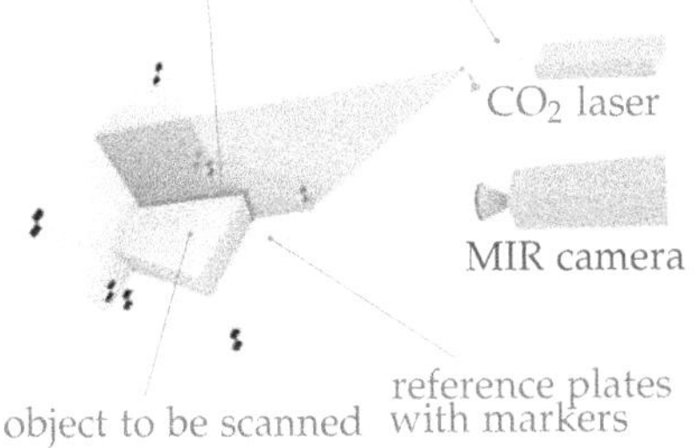

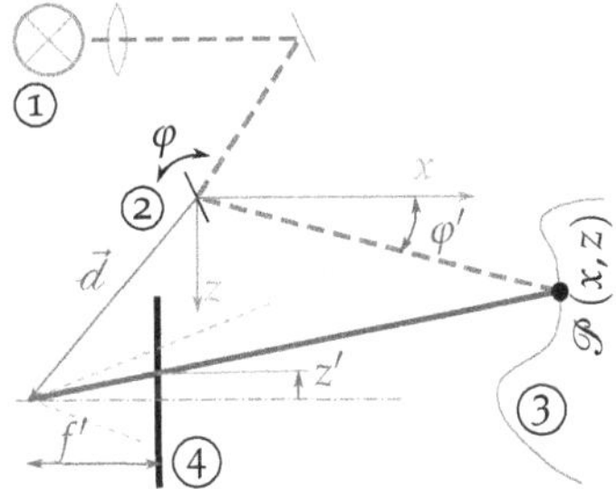

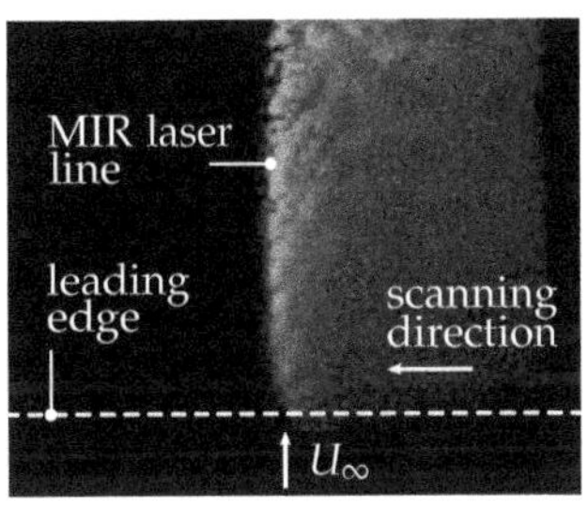

Figure 8.12: *Left:* Configuration of the laser-line scanning technique. *Centre:* Light path from laser and optics (1) that is deflected (2) on an object (3) and the triangulation principle using the image recordings (4). *Right:* MIR camera image while scanning.

configuration. A CO_2-laser beam is expanded to form a light sheet that when intersecting the ice surface is transformed into heat within a distance of a few micrometres. Consequently, the temperature increases, and the ice surface emits radiation similar to a grey body, which is recorded by a mid-infrared camera. While panning a laser mirror, the light sheet continuously scans over the entire surface.

TRIANGULATION enables surface reconstruction; see Figure 8.12 (centre). The imaging system ④ is represented by a pinhole model in this sketch and has a focal length of f'. The distance vector $\vec{d} = \binom{d_x}{d_z}$ between the panning mirror ② and imaging system remains constant. The angular position of the mirror φ determines the laser beam deflection φ'. Based on the camera coordinate z', point $\mathscr{P}(x,z)$ on the object surface can be reconstructed

$$\mathscr{P}(x,z): \quad x = \frac{d_z + d_x \frac{z'}{f'}}{\tan\varphi' + \frac{z'}{f'}} \quad , \qquad z = x \tan\varphi' \quad .$$

The unknown parameters d_x, d_z and f' are determined by a camera calibration[25] with a 3D target of known dimensions.

THE SMEARING OF THE LASER LINE PROJECTION is a notable challenge; see the example of a recorded MIR image of ice accretion[26] shown in Figure 8.12 (right). The figure shows a top view of an NACA0012 aerofoil, where the laser light sheet moves from right to left and heats up the ice surface. Note that the temperature increase is less than 1 K. Convective heat transfer by the surrounding cold airflow ultimately cools the ice surface to its initial temperature. However, this cooling process is slower than the movement of the laser line, which can thus be detected only on its leftmost edge.

[25] Franck S. Marzani, Yvon Voisin, Lew F. C. Lew Yan Voon, and Alain Diou. Calibration of a three-dimensional reconstruction system using a structured light source. *Optical Engineering*, 41(2): 484–492, 2002

[26] Xiaoliang Gong and Stephan E. Bansmer. 3-D ice shape measurements using mid-infrared laser scanning. *Optics Express*, 23(4):4908–4926, 2015a

9
Simulation of ice accretion

The testing and evaluation costs for military aircraft constitute 15 % to 20 % of the total costs, see Fox et al. (2004). Similar efforts are necessary for civil aircraft manufacturers. Although airworthiness certification still requires full-scale icing tests, design by numerical analysis has gained significant importance over recent decades. This chapter summarizes the background to computationally predict icing on aircraft.

9.1 General approach

Starting in the late 1970s, various computational approaches emerged to enable the simulation of ice accretion. Some well-established icing codes were developed by the National Aeronautics and Space Administration (NASA)[1], the French Office National d'Etudes et de Recherches Aérospatiales (ONERA)[2], the Italian Centro Italiano Ricerche Aerospaziali (CIRA)[3], the former Royal Aerospace Establishment in the UK (RAE)[4], and McGill University in Canada[5].

THE GENERAL METHODOLOGY for all these icing codes consists primarily of three distinct building blocks; see also Figure 9.1. In the first step, the airflow around the geometry of interest is computed. Within the airflow, supercooled droplets are transported that cause

[1] William Wright. Validation Results for LEWICE 3.0. Aerospace Sciences Meetings. American Institute of Aeronautics and Astronautics, January 2005. DOI: 10.2514/6.2005-1243

[2] Thierry Hedde and Didier Guffond. ONERA three-dimensional icing model. *AIAA Journal*, 33(6):1038–1045, June 1995. DOI: 10.2514/3.12795

[3] Giuseppe Mingione and Vincenzo Brandi. Ice Accretion Prediction on Multielement Airfoils. *Journal of Aircraft*, 35(2):240–246, March 1998. DOI: 10.2514/2.2290

[4] Roger W. Gent. TRAJICE2: a combined water droplet and ice accretion prediction code for aerofoils. Technical Report RAE Technical Report 90054, Royal Aerospace Establishment, 1994

[5] Héloïse Beaugendre, François Morency, and Wagdi G. Habashi. FENSAP-ICE's Three-Dimensional In-Flight Ice Accretion Module: ICE3D. *Journal of Aircraft*, 40(2):239–247, March 2003. DOI: 10.2514/2.3113

task	methods	reference
airflow solution	Euler or panel method	Hess (1990)
		Jameson, Schmidt and Turkel (1981)
	+ boundary layer method	Patankar and Spalding (1972)
	RANS method	Hirsch (2007)
droplet trajectories	cylinder approximation	Langmuir and Blodgett (1946)
		Finstad, Lozowski and Gates (1988)
	Lagrangian method	
	Eulerian method	Bourgault et al. (1999)
mass & energy balance	empirical	Makkonen (2000)
	Messinger model	Messinger (1953)
	morphogenetic approach	Szilder (1994)

Table 9.1: Different methodologies for computing ice accretion.

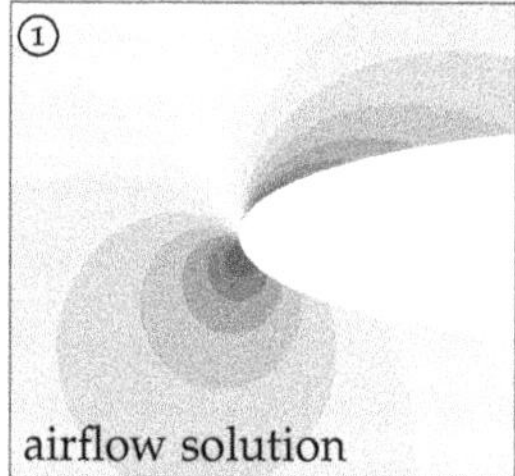

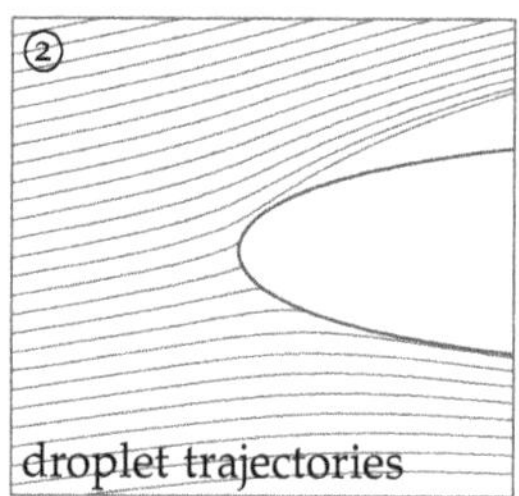

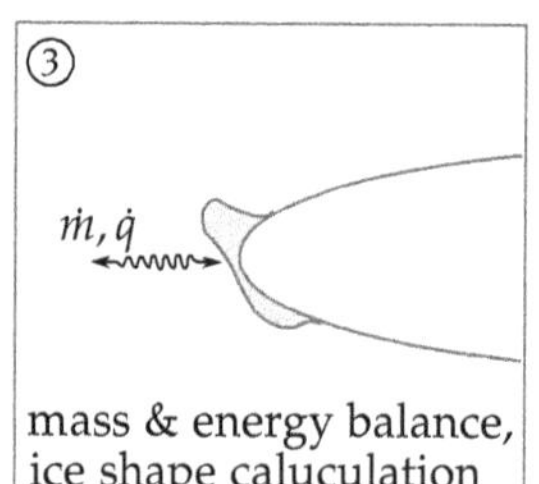

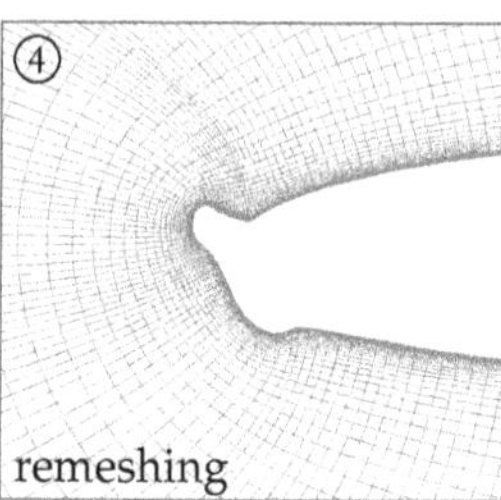

Figure 9.1: Iterative approach for simulating ice accretion with TAUICE. After calculating an ice shape in step 3, the new geometry is remeshed, and the cycle continues at step 1.

ice accretion upon impact. Therefore, the trajectories of the supercooled water droplets are determined in the next step. These trajectories are used to calculate the mass distribution of the water impinging on the geometry. Due to the low concentration of water droplets, two-way coupling between the airflow and droplet trajectory solver is not necessary. Finally, a mass and heat/energy balance is exercised to determine the local amount of freezing water. A variety of approaches and models are available to carry out these three tasks and are summarized in Table 9.1.

THROUGHOUT THIS MONOGRAPH, two-dimensional icing computations are mostly performed with the TAUICE solver, which was developed at the German Aerospace Centre (DLR) and is based on the RANS solver TAU.[6] A Lagrangian method is used to calculate particle trajectories, and a Messinger model is applied to determine ice shapes. TAUICE applies a multishot approach by splitting the ice accumulation time t_{acc} in several intervals. After one interval, the iced geometry is remeshed, and the cycle restarts with the next airflow computation; see again Figure 9.1.

[6] Dieter Schwamborn, Thomas Gerhold, and Ralf Heinrich. The DLR TAU-code: Recent applications in research and industry. In *ECCOMAS CFD 2006 conference*, 2006

9.2 *Airflow solver*

Governing equations

Compressible fluid motion is governed by the conservation of mass, momentum and energy, yielding the following transport equations:

$$\begin{aligned}
\frac{\partial \varrho}{\partial t} &+ \nabla\cdot(\varrho\vec{u}) &&= 0 \\
\frac{\partial(\varrho\vec{u})}{\partial t} &+ \nabla\cdot(\varrho\vec{u}\otimes\vec{u}) &&= -\nabla p + \nabla\cdot\boldsymbol{\tau} \\
\frac{\partial(\varrho E)}{\partial t} &+ \nabla\cdot(\varrho\vec{u}H) &&= \nabla\cdot(\boldsymbol{\tau}\cdot\vec{u}) + \nabla\cdot(k\nabla T) \quad ,
\end{aligned} \tag{9.1}$$

where ϱ is the density, $\vec{u} = (u, v, w,)^T$ the velocity, p the pressure, k the thermal conductivity coefficient and T the static temperature. The total energy E and total enthalpy H can be written as

$$E = \mathrm{e} + \frac{1}{2}|\vec{u}|^2 \qquad H = \mathrm{h} + \frac{1}{2}|\vec{u}|^2 \quad ,$$

with the specific internal energy e and the specific enthalpy h. The viscous stress tensor $\boldsymbol{\tau}$ given by

$$\boldsymbol{\tau} = \mu\left(\nabla \otimes \vec{u} + (\nabla \otimes \vec{u})^{\mathrm{T}}\right) - \frac{2}{3}\mu\,(\nabla \cdot \vec{u})\,\boldsymbol{I} \quad ,$$

is composed of two parts. The left term corresponds to NEWTON'S observation of proportionality between the shear stress and velocity gradient. The divergence $\nabla \cdot \vec{u}$ in the right term can be transformed with the continuity equation into $-\frac{1}{\varrho}\left(\frac{\partial \varrho}{\partial t} + \vec{u} \cdot \nabla \varrho\right)$. Thus, the shear stress also originates from volumetric dilation.[7]

[7] Morton E. Gurtin, Eliot Fried, and Lallit Anand. *The mechanics and thermodynamics of continua*. Cambridge University Press, 2013. ISBN 978-0521405980

TURBULENT FLOW, which is characterized by three-dimensional stochastic fluid motion, diffusion and dissipation,[8] is a common feature along aerofoils with ice accretion. When the detailed structure of turbulence is not of primary interest, the instantaneous flow equations (9.1) can be averaged, as introduced by REYNOLDS.[9] Accordingly, for any representative flow variable ϕ, a time average $\overline{\phi}$ will be defined:

[8] Stephen B. Pope. *Turbulent flows*. Cambridge Univ. Press, 12. print. edition, 2015. ISBN 9780521598866

[9] Osborne Reynolds. On the dynamical theory of incompressible viscous fluids and the determination of the criterion. *Philosophical Transactions of the Royal Society of London. A*, 186:123–164, 1895

$$\overline{\phi} = \lim_{\Delta t \to \Delta t^*} \frac{1}{\Delta t} \int_t^{t+\Delta t} \phi\,(t)\,\mathrm{d}t \quad .$$

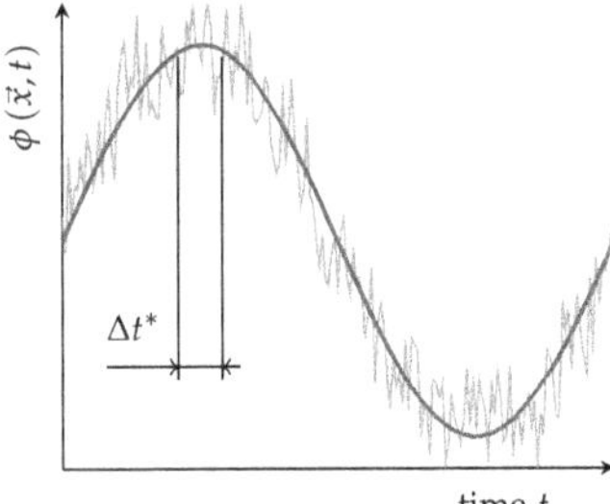

Figure 9.2: Time averaging of turbulent flow.

Note that the choice of the time interval Δt^* is not trivial; see Figure 9.2. The small-scale fluctuations are averaged while preserving the dominant unsteady modes. Moreover, the equations (9.1) incorporate compressibility effects, i.e., $\varrho = \varrho\,(p)$. It is therefore an elegant approach to apply a density-weighted averaging procedure as suggested by FAVRE[10]

[10] Alexandre Favre. Equations des gaz turbulents compressibles. *Journal de mecanique*, 4(4):391, 1965

$$\widetilde{\phi} = \frac{1}{\overline{\varrho}} \lim_{\Delta t \to \Delta t^*} \frac{1}{\Delta t} \int_t^{t+\Delta t} \varrho\,(t) \cdot \phi\,(t)\,\mathrm{d}t \quad .$$

To average the equations (9.1), the various flow properties are decomposed into a mean and a fluctuating part:

$$\begin{array}{lll} \varrho = \overline{\varrho} + \varrho' & \mathrm{e} = \widetilde{\mathrm{e}} + \mathrm{e}'' & \vec{u} = \widetilde{\vec{u}} + \vec{u}'' \\ p = \overline{p} + p' & \mathrm{h} = \widetilde{\mathrm{h}} + \mathrm{h}'' & \boldsymbol{\tau} = \widetilde{\boldsymbol{\tau}} + \boldsymbol{\tau}'' \quad . \\ & T = \widetilde{T} + T'' & \end{array}$$

When substituting the above decomposition into equations (9.1), the subsequent Favre average results in

$$
\begin{aligned}
\frac{\partial \overline{\varrho}}{\partial t} + \nabla\cdot\left(\overline{\varrho}\widetilde{\vec{u}}\right) &= 0 \\
\frac{\partial\left(\overline{\varrho}\widetilde{\vec{u}}\right)}{\partial t} + \nabla\cdot\left(\overline{\varrho}\widetilde{\vec{u}}\otimes\widetilde{\vec{u}}\right) &= -\nabla\overline{p} + \nabla\cdot\left(\widetilde{\boldsymbol{\tau}} - \overline{\varrho}\cdot\widetilde{\vec{u}''\otimes\vec{u}''}\right) \\
\frac{\partial\left(\overline{\varrho}\widetilde{E}\right)}{\partial t} + \nabla\cdot\left(\overline{\varrho}\widetilde{\vec{u}}\widetilde{H}\right) &= \nabla\cdot\left[\left(\widetilde{\boldsymbol{\tau}} - \overline{\varrho}\cdot\widetilde{\vec{u}''\otimes\vec{u}''}\right)\widetilde{\vec{u}}\right] + \nabla\cdot\left[\left(k+k_T\right)\nabla\widetilde{T}\right] + \underbrace{\nabla\cdot\left(\widetilde{\boldsymbol{\tau}\cdot\vec{u}''} - \overline{\varrho}\,\widetilde{\vec{u}''\cdot K}\right)}_{\boxtimes} \; .
\end{aligned}
\tag{9.2}
$$

These equations are very similar to (9.1), and the major differences are highlighted in blue. Observe that the viscous stress tensor $\widetilde{\boldsymbol{\tau}}$ is complemented by the Favre-averaged Reynolds stress tensor $-\overline{\varrho}\cdot\widetilde{\vec{u}''\otimes\vec{u}''}$. The Boussinesq approximation[11] suggests a further similarity between the Reynolds stress and viscous stress.

[11] Jean Boussinesq. *Essai sur la théorie des eaux courantes*. Imprimerie nationale, 1877

$$
\begin{aligned}
-\overline{\varrho}\cdot\widetilde{\vec{u}''\otimes\vec{u}''} &= \mu_T\left(\nabla\otimes\widetilde{\vec{u}} + \left(\nabla\otimes\widetilde{\vec{u}}\right)^{\mathrm{T}}\right) - \frac{2}{3}\mu_T\left(\nabla\cdot\widetilde{\vec{u}}\right)\boldsymbol{I} - \frac{2}{3}\overline{\varrho}\widetilde{K}\boldsymbol{I} \\
&= \frac{\mu_T}{\mu}\widetilde{\boldsymbol{\tau}} - \frac{2}{3}\overline{\varrho}\widetilde{K}\boldsymbol{I}
\end{aligned}
$$

The turbulent eddy viscosity μ_T is used as a scaling factor. The supplement $-\frac{2}{3}\overline{\varrho}\widetilde{K}\boldsymbol{I}$, which is a function of the turbulent kinetic energy $\widetilde{K} = \frac{1}{2}\widetilde{\vec{u}''\cdot\vec{u}''}$, satisfies the proper trace of the Reynolds stress tensor. The turbulent thermal conductivity coefficient k_T in the energy conservation equation emanates from the Reynolds analogy between momentum and heat transfer[12] and can thus be expressed as a function of the turbulent eddy viscosity, $k_T = \mu_T\cdot c_p/\,\mathrm{Pr}_T$. Herein, $\mathrm{Pr}_T = 0.9$ denotes the turbulent Prandtl number for air, and c_p is the specific heat coefficient at constant pressure. Applying the same analogy, k can be formulated as $k = \mu\cdot c_p/\,\mathrm{Pr}$. The laminar Prandtl number for air is $\mathrm{Pr} = 0.72$. The last term ⊠ in the energy equation describes the molecular diffusion and turbulent transport of the turbulent kinetic energy and is estimated as a function of μ_T and a related Prandtl number Pr_k. With the above closure approximations, the Favre-averaged conservation equations can be rewritten as

[12] Osborne Reynolds. On the Extent and Action of the Heating Surface of Stream Boilers. *Proc. Lit. Soc. Manchester*, 14: 7–12, 1874

$$
\begin{aligned}
\frac{\partial \overline{\varrho}}{\partial t} + \nabla\cdot\left(\overline{\varrho}\widetilde{\vec{u}}\right) &= 0 \\
\frac{\partial\left(\overline{\varrho}\widetilde{\vec{u}}\right)}{\partial t} + \nabla\cdot\left(\overline{\varrho}\widetilde{\vec{u}}\otimes\widetilde{\vec{u}}\right) &= -\nabla\left(\overline{p} + \frac{2}{3}\overline{\varrho}\widetilde{K}\right) + \left(1 + \frac{\mu_T}{\mu}\right)\nabla\cdot\widetilde{\boldsymbol{\tau}} \\
\frac{\partial\left(\overline{\varrho}\widetilde{E}\right)}{\partial t} + \nabla\cdot\left(\overline{\varrho}\widetilde{\vec{u}}\widetilde{H}\right) &= \nabla\cdot\left[\left(1 + \frac{\mu_T}{\mu}\right)\widetilde{\boldsymbol{\tau}}\,\widetilde{\vec{u}}\right] + \nabla\cdot\left[\left(\frac{c_p}{\mathrm{Pr}}\mu + \frac{c_p}{\mathrm{Pr}_T}\mu_T\right)\nabla\widetilde{T}\right] + \nabla\cdot\left[\left(\mu + \frac{\mu_T}{\mathrm{Pr}_k}\right)\nabla\widetilde{K}\right] \; ,
\end{aligned}
\tag{9.3}
$$

with the relations for the Favre-averaged total energy and total enthalpy:

$$\widetilde{E} = \widetilde{e} + \frac{1}{2}\left|\widetilde{\vec{u}}\right|^2 + \widetilde{K} \qquad \widetilde{H} = \widetilde{h} + \frac{1}{2}\left|\widetilde{\vec{u}}\right|^2 + \widetilde{K} = \widetilde{E} + \frac{\overline{p}}{\overline{\varrho}} \quad .$$

The Sutherland law is used to describe the temperature dependence of the dynamic viscosity:

$$\mu = \mu\left(\widetilde{T}\right) = \mu_{\text{ref}} \cdot \left(\frac{\widetilde{T}}{T_{\text{ref}}}\right)^{\frac{3}{2}} \frac{T_{\text{ref}} + 110.4\text{K}}{\widetilde{T} + 110.4\text{K}} \quad .$$

Be aware that the primary unknowns of the conservation equations (9.3) are $\overline{\varrho}$, $\widetilde{\vec{u}}$ and $\widetilde{E}$. The majority of the remaining variables can be deduced from the primary unknowns. In particular, both the pressure and temperature are obtained by the laws for an ideal gas

$$\overline{p} = \left(\frac{c_p}{c_v} - 1\right)\overline{\varrho}\left(\widetilde{E} - \frac{1}{2}\left|\widetilde{\vec{u}}\right|^2 - \widetilde{K}\right)$$

$$\widetilde{T} = \frac{1}{c_v}\left(\widetilde{E} - \frac{1}{2}\left|\widetilde{\vec{u}}\right|^2 - \widetilde{K}\right) \quad ,$$

with the specific heat coefficient at constant volume c_v. To guarantee a unique solution of the equation system (9.3), the turbulent eddy viscosity μ_T and the turbulent kinetic energy $\widetilde{K}$ have to be computed with a turbulence model.[13] For this purpose, the *negative Spalart-Allmaras* model[14] was chosen because of its simplicity and robustness. It solves a transport equation for a modified kinematic eddy viscosity $\check{\nu}_T$

$$\frac{\partial\left(\overline{\varrho}\check{\nu}_T\right)}{\partial t} + \nabla \cdot \left(\overline{\varrho}\widetilde{\vec{u}}\check{\nu}_T\right) = \text{Prod}\left(\check{\nu}_T\right) - \text{Diff}\left(\nabla^2 \check{\nu}_T\right) - \text{Dest}\left(\check{\nu}_T^2\right) \quad ,$$

and includes the production, diffusion and destruction of this quantity on the right-hand side. These terms are based on semi-empirical considerations and are calibrated at typical aeronautical flow scenarios. Free shear flows are used to adjust the production and diffusion, and the destruction arises from wall-bounded flows at high Reynolds numbers. Moreover, experimental evidence of the turbulent eddy viscosity μ_T in zero-pressure gradient boundary layers at low Reynolds numbers is taken into account; see Figure 9.3. In the viscous sublayer and buffer layer, μ_T does not exhibit the same linear behaviour $\mu_T^+ \sim z^+$ observed in the log-law region. Therefore, a modified kinematic eddy viscosity $\check{\nu}_T$ is derived that shows proportionality with the wall distance z^+ over the entire region of the inner boundary layer:

$$\check{\nu}_T = f\left(\mu_T, \overline{\varrho}, k_s, z^+\right) \quad ,$$

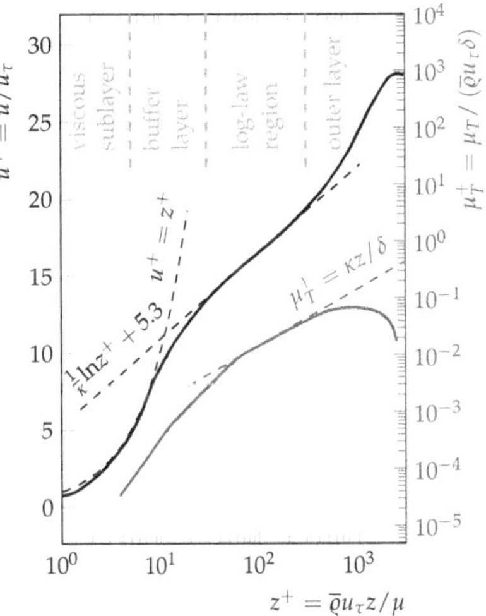

Figure 9.3: The structure of a turbulent zero-pressure gradient boundary layer and its distribution of the dimensionless turbulent eddy viscosity μ_T^+. The shear velocity $u_\tau = \sqrt{\tau_w/\overline{\varrho}}$ is a function of the wall shear stress τ_w. δ denotes the boundary layer thickness, and $\kappa \approx 0.4$ is the von Kármán constant. The data are compiled from Bradshaw (1978), Schubauer (1954), Klebanoff (1954) and Townsend (1951).

[13] David C. Wilcox. *Turbulence Modeling for CFD*, volume 2. DCW Industries, third edition edition, 2006. ISBN 978-1928729082

[14] Philippe R. Spalart and Steven R. Allmaras. A One-Equation Turbulence Model for Aerodynamic Flows. In *AIAA 30th Aerospace Sciences Meeting and Exhibit, Reno, Nevada*, number AIAA-92-0439, January 1992; and Steven R. Allmaras, Forrester T. Johnson, and Philippe R. Spalart. Modifications and Clarifications for the Implementation of the Spalart-Allmaras Turbulence Model. In *Seventh International Conference on Computational Fluid Dynamics (ICCFD7), Big Island, Hawaii*, number ICCFD7-1902, July 2012

thus simplifying the requirements of mesh resolution in the near-wall region. The effect of the equivalent sand roughness k_s is incorporated similarly. Finally, it shall be emphasized that the Spalart-Allmaras model does neglect the turbulent kinetic energy, i.e.,

$$\widetilde{K} = 0 \quad .$$

Compact notation of the governing equations

When integrating the partial differential equations for the conservation of mass, momentum and energy (9.3) over a finite control volume $V(t)$ and its boundary surface $S(t)$, the integral equation

$$\frac{\partial}{\partial t}\int_{V(t)} \underbrace{\begin{pmatrix} \overline{\varrho} \\ \overline{\varrho}\widetilde{\vec{u}} \\ \overline{\varrho}\widetilde{E} \end{pmatrix}}_{\vec{W}} \mathrm{d}V = -\int_{S(t)} \left[\underbrace{\begin{pmatrix} \overline{\varrho}\widetilde{\vec{u}} \\ \overline{\varrho}\widetilde{\vec{u}} \otimes \widetilde{\vec{u}} + \overline{p}\boldsymbol{I} \\ \overline{\varrho}\widetilde{\vec{u}}\widetilde{H} \end{pmatrix}}_{\boldsymbol{F}^{\text{conv}}} - \underbrace{\begin{pmatrix} 0 \\ \left(1 + \frac{\mu_T}{\mu}\right)\widetilde{\boldsymbol{\tau}} \\ \left(1 + \frac{\mu_T}{\mu}\right)\widetilde{\boldsymbol{\tau}}\widetilde{\vec{u}} + \left(\frac{c_p}{\mathrm{Pr}}\mu + \frac{c_p}{\mathrm{Pr}_T}\mu_T\right)\nabla\widetilde{T} \end{pmatrix}}_{\boldsymbol{F}^{\text{visc}}} \right] \cdot \vec{n}\, \mathrm{d}S$$

is obtained.[15] The primary unknowns $\left(\overline{\varrho}, \widetilde{\vec{u}}, \widetilde{E}\right)$ are grouped into one vector $\vec{W}$, which is called the *vector of the conservative variables*. If the volume integral of $\vec{W}$ changes over time, a flux over the surface S with the normal vector $\vec{n}$ pointing outwards will be entailed. The flux tensor is composed of two parts: a convective flux $\boldsymbol{F}^{\text{conv}}$ and a viscous flux $\boldsymbol{F}^{\text{visc}}$. This yields the compact notation

$$\frac{\partial}{\partial t}\int_{V(t)} \vec{W}\, \mathrm{d}V = -\int_{S(t)} \left(\boldsymbol{F}^{\text{conv}} - \boldsymbol{F}^{\text{visc}}\right) \cdot \vec{n}\, \mathrm{d}S \quad , \qquad (9.4)$$

which is advantageous for a concise description of the computational approach.

[15] Charles Hirsch. *Numerical computation of internal and external flows.* Elsevier/Butterworth-Heinemann, Amsterdam, 2007

Finite Volume Approach

The finite volume method was introduced in the field of computational fluid dynamics in the early 1970s.[16] Its foundation relies on the discretization of the integral form of the conservation equations (9.4). Therefore, the space domain is subdivided into finite control volumes V_j. The concept of median-dual control volumes is employed, as shown in Figure 9.4 for a two-dimensional illustration; a primary grid, which has been built by a mesh generator, consists of triangular and quadrilateral cells. Their vertices $\mathscr{P}_j$ will become the storage locations of the discrete vector of the conservative variables $\vec{W}_j$. Using the medians and barycentres of the primary grid cells, a polyhedral hull is established around each vertex $\mathscr{P}_j$ to form the boundary – *the faces* – of the control volume. The control volume around $\mathscr{P}_1$ has five

[16] Robert MacCormack and Alvin Paullay. Computational efficiency achieved by time splitting of finite difference operators. Aerospace Sciences Meetings. American Institute of Aeronautics and Astronautics, January 1972

faces: $\mathscr{P}_1 \rightarrow \mathscr{P}_{2..6}$. Along these faces, the fluxes of equation (9.4) must be evaluated. When using a scheme of second order in space, the face area and face-normal vectors can be simplified as

$$\text{e.g. face } \mathscr{P}_1 \rightarrow \mathscr{P}_2 : \quad \vec{n}_{1\rightarrow 2} = \frac{\vec{n}_{1\rightarrow 2,a} + \vec{n}_{1\rightarrow 2,b}}{\left|\vec{n}_{1\rightarrow 2,a} + \vec{n}_{1\rightarrow 2,b}\right|}$$

$$S_{1\rightarrow 2} = S_{1\rightarrow 2,a} + S_{1\rightarrow 2,b} \quad .$$

Further assuming average values of $\vec{W}_j$ in each control volume, the continuous integral form of the conservation equations (9.4) can be transformed into a semi-discrete one:

$$\frac{\partial}{\partial t}\left[V_j \cdot \vec{W}_j\right] = -\sum_{\text{faces of } V_j}\left[\left(\boldsymbol{F}^{\text{conv}} - \boldsymbol{F}^{\text{visc}}\right) \cdot \vec{n} \cdot S\right] \quad , \qquad (9.5)$$

which maintains the temporal derivative $\frac{\partial}{\partial t}$. Note that the *method of lines*[17] is implicitly involved in this procedure, thus permitting separate discretization in space and in time.

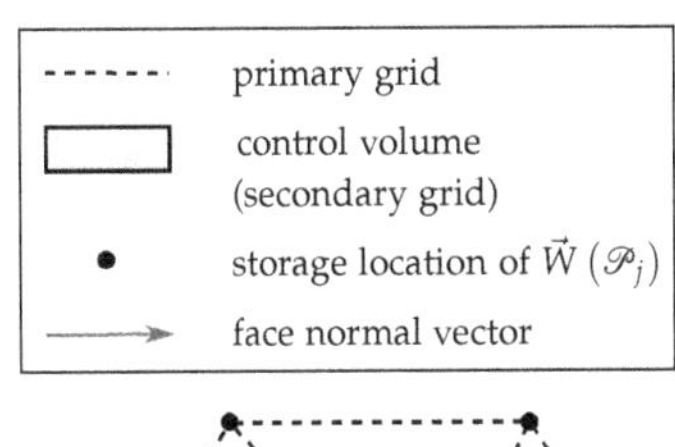

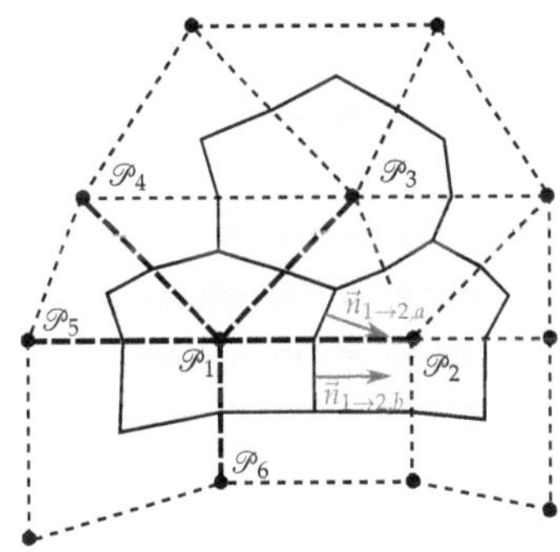

Figure 9.4: Discretization of the space domain using median-dual control volumes.

[17] William E. Schiesser. *The Numerical Method of Lines*. Elsevier Science, 2012. ISBN 9780126241303

Discretization of convective flux using a central scheme

The convective flux tensor $\boldsymbol{F}^{\text{conv}}$ of equation (9.4) can be evaluated at a certain position in space once the vector of the conservative variables $\vec{W}$ is known at the exact location. When using the discretization approach of the previous section (9.5), a non-trivial problem will arise: $\vec{W}_j$ is stored in the centre of the control volume V_j, but the convective flux tensor $\boldsymbol{F}^{\text{conv}}$ must be evaluated at its faces. This will be accomplished using the average:

$$\boldsymbol{F}^{\text{conv}}\Big|_{\text{face } \mathscr{P}_j \rightarrow \mathscr{P}_k} = \frac{1}{2}\left[\boldsymbol{F}^{\text{conv}}\left(\vec{W}_j\right) + \boldsymbol{F}^{\text{conv}}\left(\vec{W}_k\right)\right] \quad , \qquad (9.6)$$

which constitutes a *central scheme*. Let us elucidate this flux evaluation with a one-dimensional example; see Figure 9.5. Applying the semi-discretized conservation equation (9.5) for the control volume in the centre and neglecting the viscous flux, one obtains

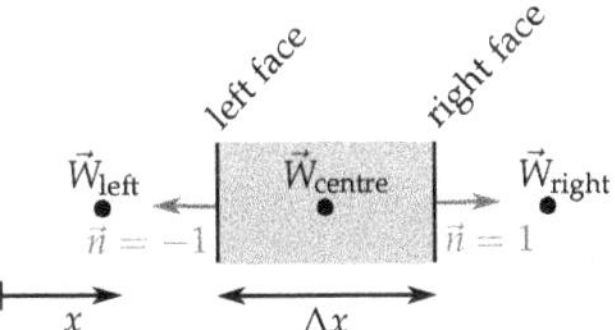

Figure 9.5: Neighbouring control volumes in a one-dimensional space and their joint faces.

$$\begin{aligned}
\frac{\partial}{\partial t}\left[V_{\text{centre}} \cdot \vec{W}_{\text{centre}}\right] &= -\left[\boldsymbol{F}^{\text{conv}}\vec{n}S\Big|_{\text{left face}} + \boldsymbol{F}^{\text{conv}}\vec{n}S\Big|_{\text{right face}}\right] \\
\frac{\partial}{\partial t}\left[\vec{W}_{\text{centre}}\right] &= -\frac{S}{V_{\text{centre}}}\left[\boldsymbol{F}^{\text{conv}}\,\vec{n}\Big|_{\text{left face}} + \boldsymbol{F}^{\text{conv}}\,\vec{n}\Big|_{\text{right face}}\right] \\
&= -\frac{1}{\Delta x}\left[-\vec{F}^{\text{conv}}_{\text{left face}} + \vec{F}^{\text{conv}}_{\text{right face}}\right] \quad .
\end{aligned}$$

Exercising the flux evaluation of (9.6)

$$\vec{F}^{\text{conv}}_{\text{left face}} = \frac{1}{2}\left[\vec{F}^{\text{conv}}\left(\vec{W}_{\text{left}}\right) + \vec{F}^{\text{conv}}\left(\vec{W}_{\text{centre}}\right)\right]$$
$$\vec{F}^{\text{conv}}_{\text{right face}} = \frac{1}{2}\left[\vec{F}^{\text{conv}}\left(\vec{W}_{\text{centre}}\right) + \vec{F}^{\text{conv}}\left(\vec{W}_{\text{right}}\right)\right] \quad ,$$

its substitution yields

$$\frac{\partial}{\partial t}\left[\vec{W}_{\text{centre}}\right] = -\frac{1}{2\,\Delta x}\left[\vec{F}^{\text{conv}}\left(\vec{W}_{\text{right}}\right) - \vec{F}^{\text{conv}}\left(\vec{W}_{\text{left}}\right)\right] \quad .$$

Now, the central difference on the right-hand side is clearly visible. Obviously, the flux evaluation of the control volume V_{centre} does not involve $\vec{W}_{\text{centre}}$ but diverts information on the direct neighbours. This is called *odd-even decoupling* and promotes the generation of high-frequency noise. Moreover, central schemes become unstable at steep gradients, as observed previously in Figure 5.4, when the VOF transport equation was analysed.

The central scheme necessitates artificial dissipation

The susceptibility of the central schemes to computational instability can be counteracted with artificial dissipation.[18] We will replace equation (9.5)

$$\frac{\partial}{\partial t}\left[V_j \cdot \vec{W}_j\right] = -\sum_{\text{faces of } V_j}\left[\left(\mathbf{F}^{\text{conv}} \cdot \vec{n} - \vec{D} - \mathbf{F}^{\text{visc}} \cdot \vec{n}\right) \cdot S\right] \quad , \quad (9.7)$$

while introducing the dissipative operator $\vec{D}$

$$\vec{D}\bigg|_{\text{face } \mathscr{P}_j \to \mathscr{P}_k} = \frac{1}{2}\Lambda_{jk}\,\mathfrak{X}\left[\epsilon^{(2)}\left(\vec{W}_j - \vec{W}_k\right) - \epsilon^{(4)}\left(\nabla^2\vec{W}_j - \nabla^2\vec{W}_k\right)\right] \quad ,$$

which must be computed at each face of the control volume. Apparently, $\vec{D}$ is a blend of second-order differences and Laplacians. The latter are activated by the indicator function $\epsilon^{(4)}$, which is on the order of unity in smooth regions of the flow field. The Laplacians act as an artificial viscosity and suppress the high-frequency noise created by the odd-even decoupling mechanism. $\epsilon^{(2)}$ is triggered near shock waves to locally enforce the scheme of being first-order accurate. The dissipative operator $\vec{D}$ further involves a scaling with the local stretching $\mathfrak{X}$ of the computational grid and the largest eigenvalue Λ_{jk} of the convective flux Jacobian, given by

$$\Lambda_{jk} = |\vec{u} \cdot \vec{n}| + a\bigg|_{\text{face } \mathscr{P}_j \to \mathscr{P}_k} \quad ,$$

where a is the speed of sound. Note that the convective flux Jacobians in mass, momentum and energy conservation involve different

[18] Antony Jameson, Wolfgang Schmidt, and Eli Turkel. Numerical Solution of the Euler Equations by Finite Volume Methods using Runge-Kutta Time-Stepping Schemes. In *AIAA-1981-1259*, 1981

eigenvalues. In the subsonic flow relevant to aircraft icing, the absolute value of the smallest eigenvalue is usually the local flow speed $\|\vec{u}\|$. Isotropic scaling with the maximum eigenvalue Λ_{jk} can thus add too much artificial dissipation. Therefore, *matrix dissipation*[19] is used instead, which replaces the scalar Λ_{jk} by the modulus of the convective flux Jacobians. As a consequence, the artificial dissipation of each conservation equation is scaled individually, thus providing more accurate solutions in the boundary layer of aerofoils subject to ice accretion.

[19] Eli Turkel. Improving the Accuracy of Central Difference Schemes. In *11th International Conference on Numerical Methods in Fluid Dynamics*, volume 323 of *Lecture Notes in Physics*, pages 586–591. Springer, 1988

Discretization of viscous flux

Beyond the convective flux, equation (9.7) necessitates a discretization of the viscous flux tensor $\boldsymbol{F}^{\text{visc}}$, which is a function of several arguments

$$\boldsymbol{F}^{\text{visc}} = \boldsymbol{F}^{\text{visc}}\left(\mu, \mu_T, \vec{u}, \nabla \otimes \vec{u}, \nabla T\right) \quad .$$

Let ϕ represent these arguments. The elliptic nature of the viscous flux permits averaging of ϕ to evaluate $\boldsymbol{F}^{\text{visc}}$ at the faces of each control volume:

$$\boldsymbol{F}^{\text{visc}}\Big|_{\text{face }\mathscr{P}_j \to \mathscr{P}_k} = \boldsymbol{F}^{\text{visc}}\left[\frac{1}{2}\left(\phi_j + \phi_k\right)\right] \quad .$$

Temperature and velocity gradients are reconstructed with a GREEN-GAUSS approach.[20] The averaging method shows similarities to a central scheme, cf. equation (9.6), yet the arguments ϕ are averaged here and not the flux itself. Nevertheless, computational instabilities similar to those of a central scheme are observed in boundary layers, where the viscous flux becomes predominant. To damp the resulting high-frequency modes, a correction of the gradient reconstruction is suggested.[21]

[20] Michael Aftosmis, Datta Gaitonde, and Sean Tavares. Behavior of linear reconstruction techniques on unstructured meshes. *AIAA Journal*, 33(11): 2038–2049, November 1995. DOI: 10.2514/3.12945

[21] Paul I. Crumpton, Pierre Moinier, and Michael B. Giles. An unstructured algorithm for high Reynolds number flows on highly stretched grids. *Numerical methods in laminar and turbulent flow*, pages 561–572, 1997

Solving steady-state problems

Rewriting the semi-discretized governing equations (9.7) yields

$$\frac{\partial}{\partial t}\left[V_j \cdot \vec{W}_j\right] = -\vec{R}_j \quad .$$

The RHS is a nonlinear function of $\vec{W}_j$ and represents the spatial discretization. It is conflated into a single variable $\vec{R}_j$, the *residual*. The LHS involves a temporal derivative. Although the LHS vanishes for steady-state conditions, it facilitates an iterative solution process. Henceforth, the goal is to discretize the temporal derivative. For this purpose, a simplified linear multistep method[22] is applied as follows:

$$\frac{V_j \boldsymbol{I}}{\Delta t}\left(\vec{W}_j^{n+1} - \vec{W}_j^{n}\right) = -\mathfrak{b}\vec{R}_j^{n+1} - (1 - \mathfrak{b})\,\vec{R}_j^{n} \quad .$$

[22] Germund G. Dahlquist. A special stability problem for linear multistep methods. *BIT Numerical Mathematics*, 3 (1):27–43, 1963

The discrete time levels are indicated by the superscripts. n represents the current time level, which is succeeded by $n+1$. By setting $\mathfrak{b} > 0$, an implicit time integration is pursued that permits large pseudo-time steps Δt. The unknown residual $\vec{R}_j^{n+1}$ is resolved by linearizing around the current time level

$$\vec{R}_j^{n+1} = \vec{R}_j^n + \left(\frac{\partial \vec{R}}{\partial \vec{W}}\right)_j^n \left(\vec{W}_j^{n+1} - \vec{W}_j^n\right) \quad ,$$

where $\frac{\partial \vec{R}}{\partial \vec{W}}$ denotes the flux Jacobian. A substitution results in the implicit scheme

$$\underbrace{\left[\frac{V_j \boldsymbol{I}}{\Delta t} + \mathfrak{b}\left(\frac{\partial \vec{R}}{\partial \vec{W}}\right)_j^n\right]}_{\text{system matrix}} \left(\vec{W}_j^{n+1} - \vec{W}_j^n\right) = -\vec{R}_j \quad ,$$

which includes a large, sparse, and non-symmetric matrix on its LHS, the *system matrix*. These equations can be solved for $\vec{W}_j^{n+1}$ efficiently with an LU-decomposition[23] given by

[23] Antony Jameson and Eli Turkel. Implicit schemes and LU decompositions. *Mathematics of Computation*, 37(156): 385–397, 1981

$$\overbrace{\left(\boldsymbol{D} + \varkappa_{\text{rel}}\boldsymbol{L}\right)\boldsymbol{D}^{-1}\left(\boldsymbol{D} + \varkappa_{\text{rel}}\boldsymbol{U}\right)}^{\text{system matrix}} \underbrace{\underbrace{\left(\vec{W}_j^{n+1} - \vec{W}_j^n\right)}_{=\Delta\vec{W}_j^n}}_{=\Delta\vec{W}_j^{\text{intermediate}}} = -\vec{R}_j \quad .$$

The system matrix is transformed into a set of three matrices: a diagonal matrix $\boldsymbol{D}$, a strictly lower triangular matrix $\boldsymbol{L}$ and a strictly upper triangular matrix $\boldsymbol{U}$. Furthermore, an over-relaxation parameter $\varkappa_{\text{rel}}$ is introduced.[24] The inversion of the system matrix proceeds in two steps: a forward sweep that evaluates an intermediate vector $\vec{W}_j^{\text{intermediate}}$

[24] Harvard Lomax and Joseph L. Steger. Relaxation Methods in Fluid Mechanics. *Annual Review of Fluid Mechanics*, 7(1): 63–88, 1975

$$\begin{aligned} \left(\boldsymbol{D} + \varkappa_{\text{rel}}\boldsymbol{L}\right)\Delta\vec{W}_j^{\text{intermediate}} &= -\vec{R}_j \quad , \\ \Longrightarrow \quad \boldsymbol{D}\Delta\vec{W}_j^{\text{intermediate}} &= -\vec{R}_j - \varkappa_{\text{rel}}\boldsymbol{L}\Delta\vec{W}_j^{\text{intermediate}} \quad , \end{aligned} \tag{9.8}$$

which is then used in a backward sweep:

$$\begin{aligned} \boldsymbol{D}^{-1}\left(\boldsymbol{D} + \varkappa_{\text{rel}}\boldsymbol{U}\right)\Delta\vec{W}_j^n &= \Delta\vec{W}_j^{\text{intermediate}} \\ \Longrightarrow \quad \boldsymbol{D}\Delta\vec{W}_j^n &= \boldsymbol{D}\Delta\vec{W}_j^{\text{intermediate}} - \varkappa_{\text{rel}}\boldsymbol{U}\Delta\vec{W}_j^n \end{aligned} \tag{9.9}$$

to determine $\Delta\vec{W}_j^n$. The solution vector $\vec{W}_j^{n+1}$ of the new time level finally reads

$$\vec{W}_j^{n+1} = \vec{W}_j^n + \Delta\vec{W}_j^n \quad .$$

This procedure is repeated to a state of $\vec{W}_j^{n+1}$ that is invariant with further iteration and independent of the initial guess $\vec{W}_j^0$. The elegance of the LU decomposition is exemplified by a coloured matrix representation of the forward sweep, cf. equation (9.8)

$$\begin{pmatrix} \blacksquare & & & & & \\ & \blacksquare & & & & \\ & & \blacksquare & & & \\ & & & \blacksquare & & \\ & & & & \blacksquare & \\ & & & & & \blacksquare \end{pmatrix} \begin{pmatrix} \Delta\vec{W}_1 \\ \vdots \\ \Delta\vec{W}_N \end{pmatrix}^{\text{inter-mediate}} = \begin{pmatrix} \vec{R}_1 \\ \vdots \\ \vec{R}_N \end{pmatrix} - \begin{pmatrix} & & & & & \\ \blacksquare & & & & & \\ \blacksquare & \blacksquare & & & & \\ \blacksquare & \blacksquare & \blacksquare & & & \\ \blacksquare & \blacksquare & \blacksquare & \blacksquare & & \\ \blacksquare & \blacksquare & \blacksquare & \blacksquare & \blacksquare & \end{pmatrix} \begin{pmatrix} \Delta\vec{W}_1 \\ \vdots \\ \Delta\vec{W}_N \end{pmatrix}^{\text{inter-mediate}} .$$

Because of the strictly lower triangular structure of the matrix $\boldsymbol{L}$ on the RHS, the intermediate vector $\vec{W}_j^{\text{intermediate}}$ is simply obtained by $N-1$ consecutive forward substitutions, starting with the first row. Only the diagonal matrix $\boldsymbol{D}$ on the LHS must be inverted. The backward sweep in equation (9.9) is carried out in a similar way. In summary, the LU decomposition efficiently circumvents the excessive computational effort of the direct inversion of the system matrix. Nevertheless, the method relies on the construction of the $\boldsymbol{L}$, $\boldsymbol{U}$ and $\boldsymbol{D}$ matrices, which is a non-trivial problem. Implementation details for unstructured grids can be found in SHAROV *et al.* (1997).[25]

[25] Dmitri Sharov, Kazuhiro Nakahashi, Dmitri Sharov, and Kazuhiro Nakahashi. Reordering of 3-D hybrid unstructured grids for vectorized LU-SGS Navier-Stokes computations. In *13th Computational Fluid Dynamics Conference*, page 2102, 1997

9.3 *Trajectories of supercooled droplets*

Lagrangian approach and solution scheme

Based upon a steady airflow solution $\vec{W} \equiv \vec{W}_{\text{air}}$, the trajectories of supercooled droplets can be computed. Similar to the considerations already presented on page 73, a Lagrangian method is applied that tracks the motion of individual droplets. The underlying model is predicated on NEWTON's second law to establish a balance of forces acting on a single particle of diameter d. After some algebra, an ODE is obtained that describes the droplet's acceleration

$$\frac{\partial \vec{u}_d\,(\vec{x},t)}{\partial t} = \frac{c_D\,\mathrm{Re}_d}{24} \cdot \frac{18\mu_{\text{air}}}{\varrho_d d^2}\,(\vec{u}_{\text{air}} - \vec{u}_d) + \left(1 - \frac{\varrho_{\text{air}}}{\varrho_d}\right)\vec{g} \quad .$$

Herein, $\vec{u}_d$ denotes the droplet velocity, $\vec{g}$ is the gravitational acceleration, and ϱ and μ represent the density and viscosity, respectively. The product of the drag coefficient and Reynolds number, $c_D \cdot \mathrm{Re}_d$, is approximated from experimental data of LANGMUIR.[26] The time integration of the ODE is performed with a RUNGE-KUTTA method.[27]

[26] Dean T. Bowden, A. E. Gensemer, and Charles A. Skeen. Engineering summary of airframe icing technical data. Technical report, Federal Aviation Agency, 1964

[27] John R. Dormand and Peter J. Prince. A family of embedded Runge-Kutta formulae. *Journal of Computational and Applied Mathematics*, 6(1):19 – 26, 1980

Catch efficiency at the aerofoil

After release in the farfield of the computational domain, the droplets are tracked until they impact on the aerofoil; see also Figure 9.6.

Since the droplet trajectories do not coincide with the streamlines of the air flow, the impingement limits become evident. Indeed, the majority of the droplets are caught in the vicinity of the aerofoil's leading edge. To quantify the mass flow rate of the impinging droplets along the aerofoil surface, a local catch efficiency β can be determined:

$$\beta = \frac{\Delta z}{\Delta s} \quad ,$$

which will be an input for the subsequent mass balance to compute the ice accretion.

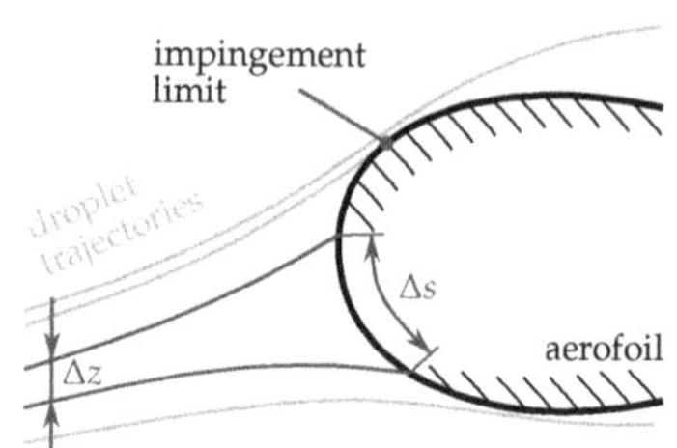

Figure 9.6: On the definition of the droplet catch efficiency β.

9.4 *Messinger model for predicting ice accretion*

To predict the amount of impinging water that will freeze on the surface of an unheated body $\mathscr{B}$, the model of MESSINGER is applied.[28,29] The model consists of a mass and energy balance of impinging, freezing, evaporating, splashing, and running water. The goal is to establish the concept of a *freezing fraction f*, which locally indicates the amount of water that will solidify to ice.

[28] Bernard L. Messinger. Equilibrium temperature of an unheated icing surface as a function of air speed. *Journal of the Aeronautical Sciences*, 20(1): 29–42, 1953

[29] Implementation of the model was kindly carried out by JAN STEINER (DLR).

THE MASS BALANCE is based on a control volume approach that is partially illustrated in Figure 9.7. The base area of the control volume is denoted as ΔS, and the time duration of the evaluated fluxes is represented by Δt. The mass flux balance reads

$$\underbrace{\dot{m}_{\text{imp}} + \dot{m}_{\text{rb,in}} + \dot{m}_{\text{st,prev}} - \dot{m}_{\text{evap}} - \dot{m}_{\text{splash}}}_{\dot{m}_{\text{avail}}} = \dot{m}_{\text{ice}} + \dot{m}_{\text{st}} + \dot{m}_{\text{rb,out}} \quad ,$$

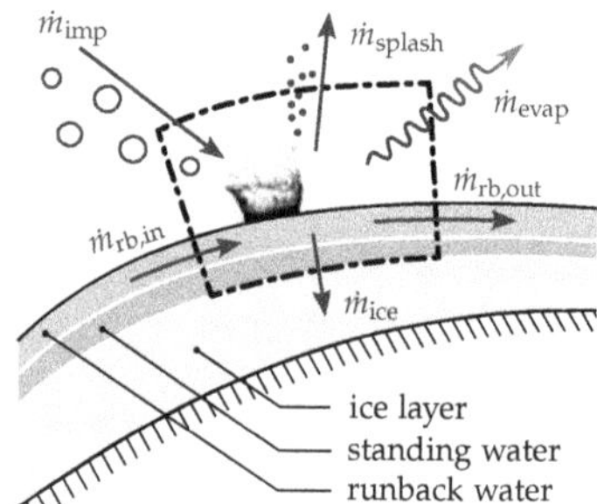

Figure 9.7: Mass balance of a control volume (dash-dot-line) along an unheated aerofoil in icing conditions.

where the individual mass contributions are explained hereafter. Note that the unit of $\dot{m}$ is $\text{kg}\,\text{s}^{-1}$. The impinging mass flow rate $\dot{m}_{\text{imp}}$ is a function of the local catch efficiency β, which results from the droplet trajectory computation outlined in the previous section. This yields

$$\dot{m}_{\text{imp}} = \beta \cdot U_\infty \cdot \text{LWC} \cdot \Delta S \quad .$$

Above a certain threshold of non-dimensional impact parameters,[30] droplet splashing must be incorporated in the mass balance. Several splashing models are presently available in industrial icing codes,[31] which again motivates the discussions in part II of this monograph. The necessary modifications can be made either directly in the droplet trajectory computation[32] or in the MESSINGER approach itself by modelling $\dot{m}_{\text{splash}}$.

[30] Damon G. K. Aboud and Anne-Marie Kietzig. Splashing threshold of oblique droplet impacts on surfaces of various wettability. *Langmuir*, 31(36): 10100–10111, 2015

[31] William Wright and Mark Potapczuk. Semi-Empirical Modelling of SLD Physics. Aerospace Sciences Meetings. American Institute of Aeronautics and Astronautics, August 2004. DOI: 10.2514/6.2004-412

[32] Raimund Honsek, Wagdi G. Habashi, and Martin S. Aubé. Eulerian modeling of in-flight icing due to supercooled large droplets. *Journal of Aircraft*, 45(4): 1290–1296, 2008

FROM THE LEFT-HAND SIDE, a thin film of liquid water – the *runback* – is entering the control volume. Continuity requires the in-flowing and out-flowing runbacks of neighbouring control volumes to be equal, i.e.,

$$\dot{m}_{\mathrm{rb,in}}\big|_i = \dot{m}_{\mathrm{rb,out}}\big|_{i-1} \quad .$$

Due to viscous effects and surface tension forces, not all of the liquid water will enter or leave the control volume as runback; some amount will remain inside the control volume as standing water, which is calculated by

$$\dot{m}_{\mathrm{st}} = \min\left(\rho_{\mathrm{W}}\frac{h_{\mathrm{rb}}\cdot\Delta S}{\Delta t},\, (1-f)\cdot\dot{m}_{\mathrm{avail}}\right) \quad ,$$

where the water film height h_{rb} is estimated by

$$h_{\mathrm{rb}} = \sqrt[3]{\left(\frac{2\,\mu_{\mathrm{W}}}{\mathfrak{w}\cdot\tau_{\mathrm{W}}}\right)^2\frac{\sigma}{\varrho_{\mathrm{W}}}} \quad .$$

Herein, τ_{W} represents the wall shear stress, and $\mathfrak{w}$ denotes an empirical wetting factor. Within the droplet impingement limits, i.e., for $\dot{m}_{\mathrm{imp}} > 0$, the wetting factor is set to $\mathfrak{w} = 1$; elsewhere, it is $\mathfrak{w} = 1/3$. Notably, the control volume balance further includes a temporal aspect; similarly, standing water from the previous time step $\dot{m}_{\mathrm{st,prev}}$ must be incorporated.

A DIFFERENCE between the water vapour pressures of the surface and boundary layer edge enables evaporation of water. The similarity between heat convection and mass exchange[33] permits the use of the heat transfer coefficient $\hbar$ for the determination of the evaporative mass flux

$$\dot{m}_{\mathrm{evap}} = \frac{\hbar}{\rho_{\mathrm{air}}\cdot c_{\mathrm{p,air}}\cdot\mathscr{R}_{\mathrm{air}}\cdot\mathrm{Le}_{\mathrm{air}}^{\frac{2}{3}}}\,\frac{M_{\mathrm{W}}}{M_{\mathrm{air}}}\left(\frac{p_{\mathrm{vap,S}}}{T_{\mathrm{S}}} - \frac{p_{\mathrm{vap,air}}\cdot\mathrm{RH}}{T_{\mathrm{air}}}\right)\Delta S \quad .$$

[33] Max Jakob. *Heat transfer*. Wiley, New York, NY, 1949

$c_{\mathrm{p,air}}$	constant-pressure specific heat of air
$\mathrm{Le}_{\mathrm{air}}$	Lewis number
M_{W}	molecular mass of water
M_{air}	molecular mass of air
$p_{\mathrm{vap,air}}$	vapour pressure of water at the boundary layer edge
$p_{\mathrm{vap,S}}$	vapour pressure of water at the surface
R_{air}	individual gas constant of air
RH	relative humidity
T_{air}	temperature at the boundary layer edge
T_{S}	surface temperature

Using the freezing fraction f, the amount of solidifying water is determined as

$$\dot{m}_{\mathrm{ice}} = f\cdot\dot{m}_{\mathrm{avail}} \quad .$$

The freezing fraction, one of the major unknowns in the equation system, can take values in the range $0 \le f \le 1$. For $f = 0$, the entire water mass will remain liquid; for $f = 1$, the entire water mass becomes ice. The second unknown is the outgoing runback, which is deduced from the mass balance:

$$\dot{m}_{\mathrm{rb,out}} = \dot{m}_{\mathrm{avail}} - \dot{m}_{\mathrm{ice}} - \dot{m}_{\mathrm{st}} \quad .$$

AN ENERGY BALANCE is hence necessary to solve for the unknowns; see Figure 9.8. The balance consists primarily of heat fluxes $\dot{q}$ with units of $\mathrm{W\,m^{-2}}$ and is given by

$$\dot{q}_{\mathrm{fus}} = \dot{q}_{\mathrm{conv,net}} + \dot{q}_{\mathrm{sens}} + \dot{q}_{\mathrm{imp}} - \dot{q}_{\mathrm{splash}} + \dot{q}_{\mathrm{cond}} + \dot{q}_{\mathrm{evap}} \quad .$$

The surrounding airflow induces a net convective heat flux $\dot{q}_{\text{conv,net}}$ composed of frictional heating $\dot{q}_{\text{air}}$ in the boundary layer and pure convection $\dot{q}_{\text{conv}}$, yielding

$$\begin{aligned}\dot{q}_{\text{conv,net}} &= \dot{q}_{\text{conv}} - \dot{q}_{\text{air}} \\ &= \hbar \cdot \mathfrak{A}_{\text{bead}} \cdot \left[\underbrace{(T_S - T_\infty)}_{\text{convection}} - \underbrace{(T_{\text{rec}} - T_\infty)}_{\text{friction}} \right] \\ &= \hbar \cdot \mathfrak{A}_{\text{bead}} \cdot (T_S - T_{\text{rec}}) \quad .\end{aligned}$$

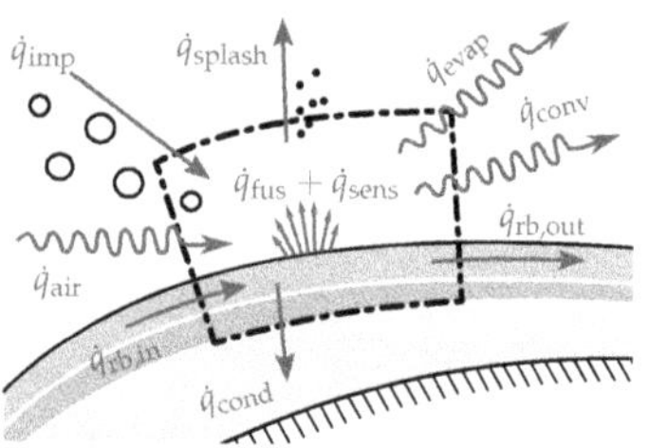

Figure 9.8: Energy balance of a control volume (dash-dot line) along an unheated aerofoil in icing conditions.

The involved static temperatures are T_S at the ice surface, T_∞ at freestream conditions and the recovery temperature T_{rec}. The unevenness of the surface structure of the ice slightly increases the effective area for convective heat transfer. The empirical factor $\mathfrak{A}_{\text{bead}}$ given by

$$\mathfrak{A}_{\text{bead}} = \frac{2}{1 + \cos\left(\frac{\pi}{90} \cdot \frac{T_{\text{m}} - T_\infty}{1\,\text{K}}\right)} \quad ,$$

accounts for that circumstance and assumes the surface to be composed of small spherical beads. The surrounding airflow is further laden with supercooled droplets. Upon impact, they cause a specific kinetic energy $\dot{q}_{\text{imp}}$, given by

$$\dot{q}_{\text{imp}} = -\frac{1}{2} \cdot \dot{m}_{\text{imp}} \cdot U_\infty^2 \cdot \frac{1}{\Delta S} \quad ,$$

where U_∞ denotes the freestream velocity. The specific kinetic energy of splashed droplets $\dot{q}_{\text{splash}}$ can be treated similarly. Because the temperature of the supercooled droplets T_∞ is usually lower than the surface temperature T_S, a sensible heat flux is generated that consists of two parts:

$$\dot{q}_{\text{sens}} = \dot{q}_{\text{sens,avail}} + \dot{q}_{\text{sens,fus}} \quad .$$

$\dot{q}_{\text{sens,avail}}$ denotes the heat loss due to heating the available water to the freezing temperature $T_{\text{m}} = 273.15\,\text{K}$,

$$\dot{q}_{\text{sens,avail}} = \dot{m}_{\text{avail}} \cdot c_{\text{p,water}} \cdot (T_{\text{m}} - T_\infty) \cdot \frac{1}{\Delta S} \quad .$$

When the surface temperature T_S is not equal to T_{m}, ice is cooled ($f = 1$) or water is heated ($f = 0$), and hence

$$\dot{q}_{\text{sens,fus}} = \begin{cases} \dot{m}_{\text{avail}} \cdot c_{\text{p,ice}} \cdot (T_S - T_{\text{m}}) \cdot \frac{1}{\Delta S} & f = 1 \\ \dot{m}_{\text{avail}} \cdot c_{\text{p,w}} \cdot (T_S - T_{\text{m}}) \cdot \frac{1}{\Delta S} & f = 0 \\ 0 & 0 < f < 1 \end{cases} \quad .$$

Moreover, the phase change of water from liquid to solid will release the latent heat of fusion $\dot{q}_{\text{fus}}$, which is determined by

$$\dot{q}_{\text{fus}} = \Delta H_{\text{fus}} \cdot f \cdot \dot{m}_{\text{avail}} \cdot \frac{1}{\Delta S} \quad ,$$

where the enthalpy of fusion is given by

$$\Delta H_{\mathrm{fus}} = c_{\mathrm{p,ice}} \left(T_{\mathrm{m}} + \Delta T_{\mathrm{res}} \right) + L_{\mathrm{fus}} - c_{\mathrm{p,ice}} T_{\mathrm{m}} \frac{\varrho_{\mathrm{ice}}}{\varrho_{\mathrm{w}}} \quad .$$

$\Delta T_{\mathrm{res}} = 10^{-5}$ K is solely included for reasons of computational stability. Evaporating water will contribute to the heat balance; likewise,

$$\dot{q}_{\mathrm{evap}} = \dot{m}_{\mathrm{evap}} \cdot L_{\mathrm{evap}} \cdot \frac{1}{\Delta S} \quad .$$

Finally, some heat is conducted into the ice surface. The corresponding heat flux is calculated by

$$\dot{q}_{\mathrm{cond}} = \sqrt{\frac{k_{\mathrm{ice}}\, \varrho_{\mathrm{ice}}\, c_{\mathrm{p,ice}}}{\pi} \cdot \frac{1}{t}} \left(T_{\mathrm{S}} - T_{\mathrm{rec}} \right) \quad ,$$

assuming the surface to be a semi-infinite slab that experiences a temperature jump $T_{\mathrm{S}} - T_{\mathrm{rec}}$ when advancing in time t.

L_{fus}	latent heat of fusion
L_{evap}	latent heat of evaporation
k_{ice}	thermal conductivity of ice

10

Understanding glaze ice accretion

Glaze ice accretion represents a transient process that can be subdivided into two major phases. During the initial stage, the wetting dynamics of the dry, clean aerofoil are prevalent. As soon as the aerofoil is covered by a thin ice layer, icing becomes an interface phenomenon of water in its different phase states that governs further ice growth. This chapter aims to summarize the different stages of ice accretion by combining theoretical, experimental and computational results. Special emphasis is placed on ice roughness because of its outstanding role in the growth dynamics of glaze ice accretion.

10.1 *Selected test cases*

For glaze ice conditions, supercooled droplets do not instantly freeze upon impact. The delicate interplay between the kinetic and thermodynamic timescales of the spreading water governs the ice accretion process. To better understand the involved phenomena, a select number of test cases have been chosen, whose boundary conditions are summarized in Table 10.1.

THE STUDY draws on two-dimensional aerofoils. NACA0012 aerofoils have been chosen because of the vast data available in the literature. The *engine nacelle* model represents a two-dimensional extrusion of an engine inlet during take-off and climb configurations.[1] Scallop icing,[2] which occurs on swept wings, is not considered in this chapter. Primarily, FAR/EASA CS25 Appendix C conditions are addressed, where the MVD is smaller than 50 µm. Supercooled large droplet (SLD) icing is covered by cases #I2 and #I5. Experimental results for #I2-#I5 are gathered in the Braunschweig Icing Wind Tunnel. The computations with TAUICE focus on case #I1.

[1] Tafara E. Makuni, Hardeep S. Kalsi, Paul G. Tucker, and Holger Babinsky. Transitional Shock-Wave/Boundary-Layer Interactions in Intakes at Incidence. In *Advances in Simulation of Wing and Nacelle Stall*, pages 323–333, 2016. DOI: 10.1007/978-3-319-21127-5_19

[2] Mario Vargas, Jen-Ching Tsao, and Alric Rothmayer. Review of Role of Icing Feathers in Ice Accretion Formation. In 2007 *SAE Aircraft and Engine Icing International Conference*. SAE International, sep 2007. DOI: https://doi.org/10.4271/2007-01-3294

Table 10.1: Boundary conditions of the test cases for analysing ice accretion phenomena that are studied in this chapter.

Reference	Case ID #	test model	chord m	U_∞ $\mathrm{m\,s^{-1}}$	T_∞ °C	LWC g/m³	MVD µm	t_{acc} s
Wright and Rutowski (1999), case 401	I1	NACA 0012	0.53	102.8	-7.8	0.55	20	420
Sommerwerk, Horst and Bansmer (2016)	I2	NACA 0012	1.0	40	-10	1.3	80	240
private data	I3	engine nacelle	0.5	40	-5	0.42	50	40
of VELANDIA	I4	engine nacelle	0.5	40	-15	0.45	40	120
Steiner and Bansmer (2016)	I5	NACA 0012	0.65	40	-20 → -5	1.3	80	44 → 737

10.2 Water collection

The collection of water is a necessary condition for ice accretion on an aerofoil. The areas that are directly facing the cloud of super-cooled droplets collect the majority of the water. Analysing the droplet trajectories allows the water collection efficiency β to be quantified. Using the algorithm presented in Section 9.3, the distribution of β is computed for test case #I1, an NACA0012 aerofoil at 3.5° AoA. Note that all of the following results will be plotted over the surface coordinate s, which has its origin at the stagnation line; see Figure 10.1.

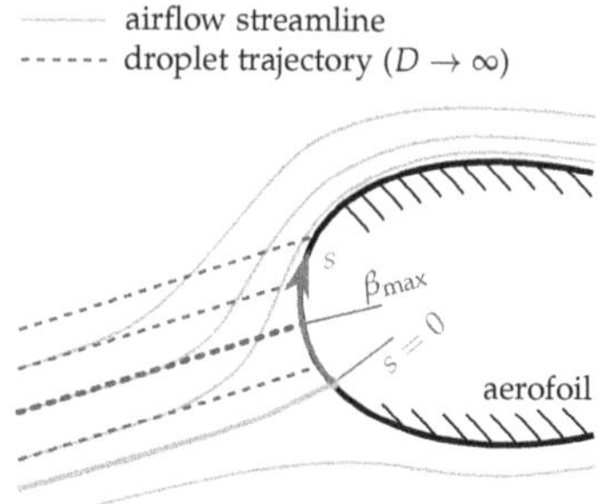

Figure 10.1: The definition of the s-coordinate and the location of β_{max}.

Very large droplets with diameter $D \to \infty$ have an excessively high inertia and thus are not influenced by the curvature of the streamlines. Neglecting gravity, their trajectories represent straight lines. In this case, the collection efficiency can be determined solely by geometrical considerations. For positive angles of attack, it is then obvious that the point of the maximum collection efficiency β_{max} is located at $s > 0$; see again Figure 10.1. Further away from the location of β_{max}, the impact angle of the droplet trajectories becomes smaller than 90°, resulting in decreased collection efficiencies.

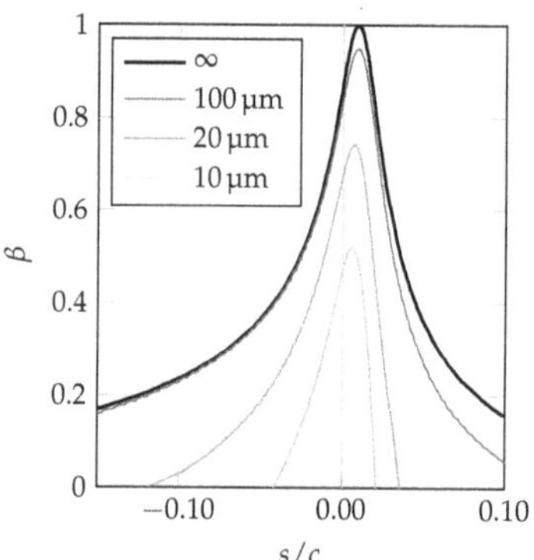

Figure 10.2: Collection efficiency β for different droplet diameters of test case #I1. Data provided by Sotomayor.

The droplet size will influence the distribution of β, as depicted in Figure 10.2. For decreasing droplet sizes, the impingement limits – where $\beta = 0$ – shift towards the aerofoil's leading edge. Smaller droplets are better able to follow the airflow streamlines. Obviously, only a tiny area of approximately $s = 0$ collects water when the droplets are only 10 µm in diameter. Importantly, ice accretion can exceed the impingement limits in the presence of glaze ice conditions. The aerodynamic shear stress at the interface between water and the surrounding air will drive the water film further downstream before it solidifies.

10.3 Initial wetting

The collection of water on the aerofoil is not a continuous process since the oncoming flow consists of individual droplets. Indeed, low LWC suggests a sequence of discrete impact events nearly following a stochastic Poisson process. Upon impact on the initially dry aerofoil, they will spread as analysed in Chapter 6, thus wetting[3] the surface.

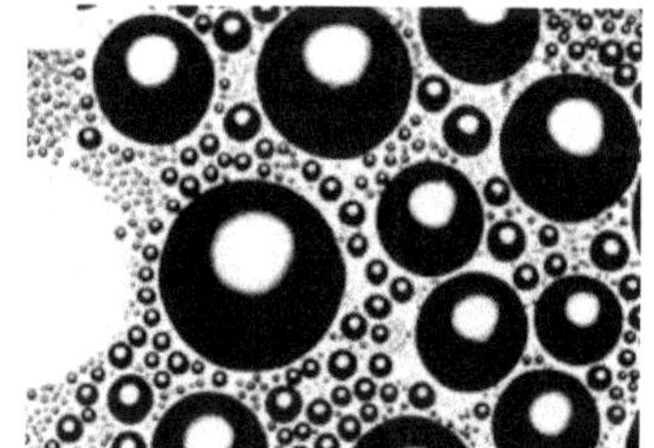

Figure 10.3: Droplet coalescence during dew formation. Reprinted from Briscoe and Galvin (1991) with permission from Elsevier.

If the solidification process is slower than the kinetic phenomena – which is usually the case for glaze ice accretion – the deposited

[3] Pierre G. de Gennes. Wetting: statics and dynamics. *Reviews of Modern Physics*, 57:827–863, Jul 1985. DOI: 10.1103/RevModPhys.57.827

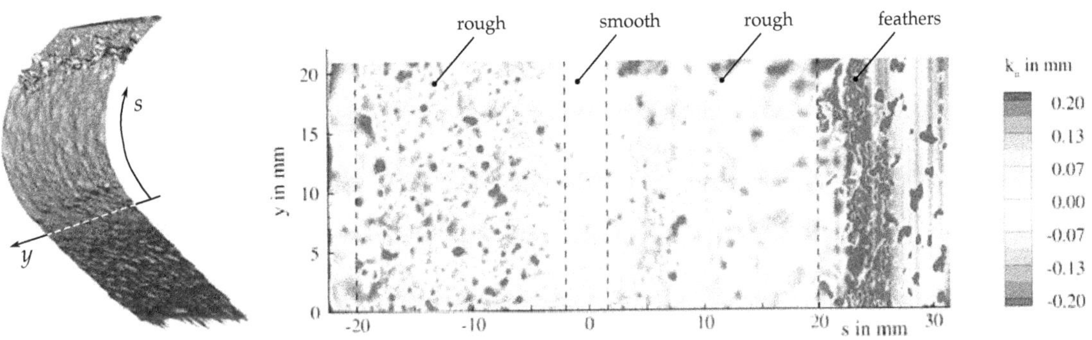

Figure 10.4: Tomographic scan of the ice accretion of test case #I3 and its deconvolved roughness distribution. Data from JUAN VELANDIA.

neighbouring droplets can coalesce. OLSEN and WALKER observed this behaviour with close-up movies of incipient ice accretion.[4] Similarities to dew formation are to be considered, i.e., air moisture condensation on cold solid surfaces. Breath figures[5,6] form, as depicted in Figure 10.3. The resulting patterns are much larger than the typical size of the impacting droplets. The largest structures are called *beads*. As soon as the beads reach a critical size, they become susceptible to the surrounding airflow and start deforming and moving.

[4] William Olsen and Ernie Walker. Experimental evidence for modifying the current physical model for ice accretion on aircraft surfaces. Technical Report NASA-TM-87184, NASA Lewis Research Center, 1986

[5] Daniela Fritter, Charles M. Knobler, and Daniel A. Beysens. Experiments and simulation of the growth of droplets on a surface (breath figures). *Physical Review A*, 43:2858–2869, Mar 1991. DOI: 10.1103/PhysRevA.43.2858

[6] Brian J. Briscoe and Kevin P. Galvin. The effect of surface fog on the transmittance of light. *Solar Energy*, 46(4):191–197, 1991. DOI: https://doi.org/10.1016/0038-092X(91)90063-3

10.4 *Initial roughness formation*

Meanwhile, solidification is triggered, and the inner core of the beads starts freezing. The process is similar to the solidification of a sessile droplet, which is discussed in Section 4.3. Note that sustained droplet impact events increase the available water mass on the aerofoil surface, hence promoting shallow water and rivulet flow around the beads.

THE INITIAL ROUGHNESS FORMATION is further studied geometrically. Therefore, a plastic replica is created from the ice accretion of test case #I3 using the mould and cast technique. X-ray microtomography with a setup similar to that described in Figure 3.5 is conducted to measure the complex surface topology of the accreted ice. Contrary to conventional imaging techniques, the tomographic approach is able to capture the ice topology in deep chasms, which would otherwise be inaccessible due to optical occlusion. Such chasms become especially relevant in the later stage of ice accretion, when ice feathers[7] form.

[7] Mario Vargas, Jen-Ching Tsao, and Alric Rothmayer. Review of Role of Icing Feathers in Ice Accretion Formation. In *2007 SAE Aircraft and Engine Icing International Conference*. SAE International, sep 2007. DOI: https://doi.org/10.4271/2007-01-3294

FIGURE 10.4 depicts the reconstructed ice surface in three-dimensional space, which is subsequently deconvolved on a y-s-plane. Four distinct zones of roughness can be identified. Close to the stagnation line ($s = 0$), the surface is very smooth. Further away on both sides, there is a sudden increase in roughness, with a structure that seems to be similar to the beading process highlighted in Figure 10.3. For $s > 20\,\mathrm{mm}$, the origin of feather growth can be observed.

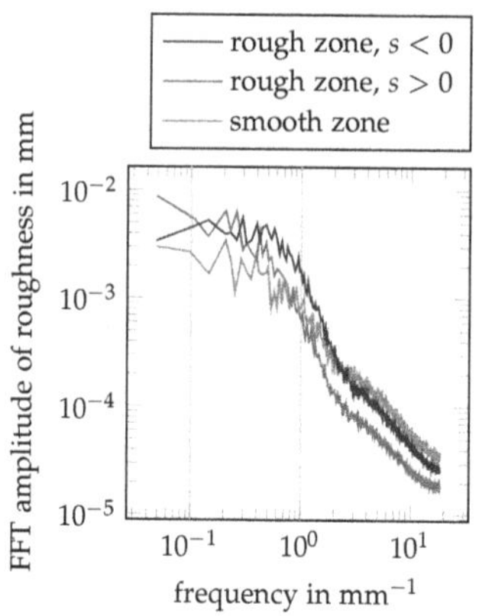

Figure 10.5: Fourier transform of the three zones of roughness of test case #I3. Data from JUAN VELANDIA.

A SPATIALLY AVERAGED TWO-DIMENSIONAL DISCRETE FOURIER TRANSFORM of the roughness distribution for each individual zone is presented in Figure 10.5. Generally, the spectrum decays monotonically with increasing frequency, and no distinct maximum is visible. Nevertheless, the frequency band from $0.1\,\mathrm{mm}^{-1}$ to $1\,\mathrm{mm}^{-1}$ is noteworthy and corresponds to wavelengths from 1 mm to 10 mm. In that interval, the roughness height of the rough zones is one order of magnitude larger than those of the smooth zone. It describes the beading process, which is indeed in the millimetre-size range.

AS TIME PROGRESSES, the roughness elements grow in size. This is visualized with a series of photographs of test case #I2, which are post-processed using a Sobel operator for edge detection, see Figure 10.6. Over time, the extent of the smooth zone close to the stagnation line decreases, and after three minutes of ice accumulation, it completely vanishes.

10.5 *Roughness and convective heat transfer*

The growing roughness elements alter the wall-near boundary-layer flow. In this section, it will be shown that the properties of the thermal boundary layer above the ice are also affected. The increased temperature gradients promote an augmented convective heat transfer that favours ice accretion.

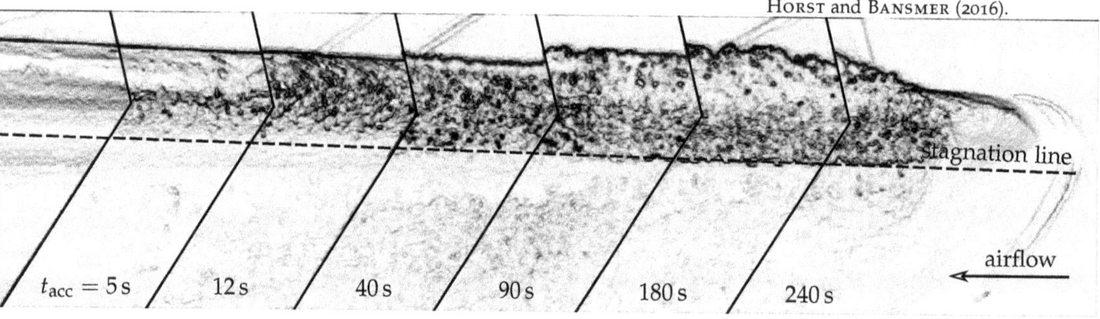

Figure 10.6: Ice accretion on an NACA0012 aerofoil over time, post-processed with a Sobel operator, cf. Figure 2.1. Test case #I2. $T_\infty = 263.15\,\mathrm{K}$, LWC $= 1.3\,\mathrm{g/m^3}$, MVD $= 80\,\mu\mathrm{m}$, AoA $= 8°$. Data from SOMMERWERK, HORST and BANSMER (2016).

Analogy of velocity and thermal boundary layer

In the boundary layer, the wall-normal gradients of the velocity and temperature are much larger than those along the wall. Moreover, the wall-parallel velocity component u is significantly larger than the wall-normal component w. Invoking these boundary layer approximations in the equations for momentum and energy conservation (9.2) yields the following set of equations:

$$\begin{aligned} \varrho\left(\overline{u}\frac{\partial\overline{u}}{\partial x}+\overline{w}\frac{\partial\overline{u}}{\partial z}\right) &= -\frac{\partial\overline{p}}{\partial x}+\frac{\partial}{\partial z}\left(\mu\frac{\partial\overline{u}}{\partial z}-\varrho\overline{u'w'}\right) \\ \varrho c_p\left(\overline{u}\frac{\partial\overline{T}}{\partial x}+\overline{w}\frac{\partial\overline{T}}{\partial z}\right) &= \frac{\partial}{\partial z}\left(k\frac{\partial\overline{T}}{\partial z}-\varrho c_p\overline{T'w'}\right) \quad . \end{aligned}$$

Note that the energy equation is expressed in terms of temperature, which is a result of extensive algebra.[8] Incompressible flow and constant values of viscosity μ, heat capacity c_p and heat conductivity k are assumed, thus simplifying the Favre averaging towards a Reynolds averaging procedure. Further suppose parallel flow ($\overline{w}=0$) in the moving equilibrium; i.e., the temperature, pressure and velocity vary sufficiently slowly with x. The equations hence reduce to

$$\begin{aligned} 0 &= \frac{\partial}{\partial z}\left(\mu\frac{\partial\overline{u}}{\partial z}-\varrho\overline{u'w'}\right) \\ 0 &= \frac{\partial}{\partial z}\left(k\frac{\partial\overline{T}}{\partial z}-\varrho c_p\overline{T'w'}\right) \quad . \end{aligned}$$

Integrating the PDEs

$$\begin{aligned} \mu\frac{\partial\overline{u}}{\partial z}-\varrho\overline{u'w'} &= \text{const} = \overline{\tau}_w \\ k\frac{\partial\overline{T}}{\partial z}-\varrho c_p\overline{T'w'} &= \text{const} = \overline{\dot{q}}_w \end{aligned} \qquad (10.1)$$

gives the wall shear stress $\overline{\tau}_w$ and the wall heat flux $\overline{\dot{q}}_w$. In turbulent flow, the wall shear stress is thus caused by viscosity and momentum transfer due to turbulent fluctuations $\overline{u'w'} < 0$. In the same sense, heat transfer is enhanced due to the correlation $\overline{T'w'}$. Nondimensionalizing the above equations with the boundary layer thickness δ as a length scale, the edge velocity $\overline{u}_e$ of the boundary layer as a velocity scale and a reference temperature T_{ref} yields

$$\begin{aligned} c_f &= \frac{2}{\text{Re}_\delta}\frac{\partial\overline{u^\diamond}}{\partial z^\diamond}-2\overline{u^{\diamond\prime}w^{\diamond\prime}} \\ \text{Stn} &= \frac{1}{\text{Re}_\delta\,\text{Pr}}\frac{\partial\overline{T^\diamond}}{\partial z^\diamond}-\overline{T^{\diamond\prime}w^{\diamond\prime}} \quad , \end{aligned}$$

where c_f denotes the skin-friction coefficient and Stn is the Stanton number, which characterizes the heat transfer. Observe the similarity

[8] Frank P. Incropera and David P. DeWitt. *Fundamentals of heat and mass transfer*. Wiley, New York, 5. ed edition, 2002. ISBN 0471386502

between the two equations. They are linked by the Prandtl number $\mathrm{Pr} = c_p \cdot \mu / k$, which constitutes an analogy between the velocity and thermal boundary layer,[9] cf. equation (9.3). To understand the relation between the roughness and heat transfer, we will thus proceed with a consideration of the velocity boundary layer.

[9] Thomas H. Chilton and Allan Philip Colburn. Mass transfer (absorption) coefficients prediction from data on heat transfer and fluid friction. *Industrial & engineering chemistry*, 26(11):1183–1187, 1934

Roughness and the turbulent velocity boundary layer

Primarily depending on the roughness height and the free-stream turbulence level, the flow in the vicinity of the initial ice roughness is usually transitional or turbulent.[10] Therefore, a turbulent boundary layer shall be considered in this section. Revisiting equation (10.1) reveals a layered structure of the turbulent boundary layer, which is described in detail in classic textbooks,[11] cf. Figure 9.3. Close to the wall, the correlation $\overline{u'w'}$ vanishes, and viscous momentum transfer predominates. The friction velocity $u_\tau = \sqrt{\tau_w / \varrho}$ becomes an important scaling parameter. It can be combined with the kinematic viscosity ν to form a viscous length scale ν / u_τ, which is approximately 1 µm to 2 µm for the investigated test case #I1. Let us define the non-dimensional groups

$$u^+ = \frac{\overline{u}}{u_\tau} \qquad z^+ = \frac{z \cdot u_\tau}{\nu} \quad .$$

With increasing wall distance ($5 < z^+ < 40$), the turbulent fluctuations increase. Low-speed fluid is violently ejected outwards, and high-speed fluid rushes in towards the wall.[12] For $z^+ > 40$, the viscous stress in equation (10.1) can be neglected. An asymptotic matching analysis[13] yields the *logarithmic overlap law*, given by

$$\lim_{z^+ \to \infty} u^+(z^+) = \frac{1}{\kappa} \ln z^+ + C \quad ,$$

where $\kappa \approx 0.4$ denotes the von Kármán constant and $C \approx 5.1$ is determined experimentally for smooth walls.

[10] Michael Bragg, Matthew Cummings, and Chad Henze. Boundary-layer and heat-transfer measurements on an airfoil with simulated ice roughness. In *34th Aerospace Sciences Meeting and Exhibit*, number AIAA-96-0866 in Aerospace Sciences Meetings. American Institute of Aeronautics and Astronautics, August 1996. DOI: 10.2514/6.1996-866

[11] Peter Bradshaw, editor. *Turbulence*. Number 12 in Topics in Applied Physics. Springer, Berlin, 2., corr. and updated ed edition, 1978. ISBN 0387088644

[12] Stephen K. Robinson. Coherent Motions in the Turbulent Boundary Layer. *Annual Review of Fluid Mechanics*, 23(1):601–639, 1991

[13] Hermann Schlichting. *Boundary-Layer theory*. Springer, Berlin, 9th edition, 2017. ISBN 9783662529171

Roughness interferes with the structure of the velocity boundary layer.[14] Reconsider the measured ice roughness k of test case #I2 in the range from 100 µm to 500 µm. This corresponds to viscous scales of $k^+ = k \cdot \nu / u_\tau = 100 \cdots 500$, thus affecting all regions of the inner boundary layer. As a result, the mean velocity profile near the wall is altered; see Figure 10.7 (left). With increasing sand-grain roughness k_s^+, the universal velocity profile $u^+(z^+)$ is displaced further downward. The significance of the term "sand-grain roughness" dates back to early experiments by Nikuradse.[15] He studied pipe flow while systematically varying the wall roughness with closely packed

[14] Javier Jiménez. Turbulent flows over rough walls. *Annual Review of Fluid Mechanics*, 36:173–196, 2004; and Michael R. Raupach, Robert A. Antonia, and Sundara Rajagopalan. Rough-wall turbulent boundary layers. *Applied Mechanics Reviews*, 44(1):1–25, 1991

[15] Johann Nikuradse. Strömungsgesetze in rauhen Rohren. *VDI Forschungsheft*, 361, 1933

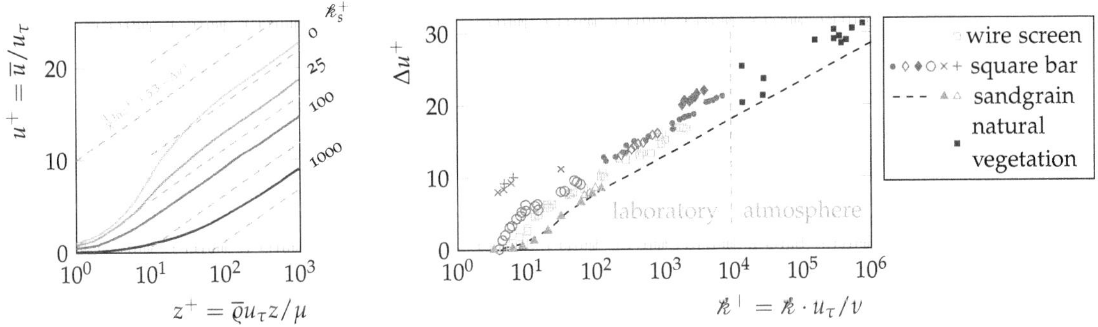

Figure 10.7: Influence of roughness on the turbulent boundary layer. *Left:* Universal velocity profiles of the inner boundary layer depending on the surface sand-grain roughness k_s^+. Adapted from ROTTA (1950). *Right:* Relationship between Δu^+ and the roughness k^+. Reproduced from RAUPACH et al. (1991) with permission from ASME.

sand. His investigations revealed that the logarithmic overlap law is also valid for rough walls:

$$\lim_{z^+ \to \infty} u^+(z^+) = \frac{1}{\kappa} \ln z^+ + C - \Delta u^+ \quad .$$

Therefore, the velocity slopes in Figure 10.7 (left) approaches asymptotically towards $1/\kappa \cdot \ln z^+$. The shift values Δu^+ closely correlate with the roughness height k^+, as highlighted in Figure 10.7 (right).

THE SKIN FRICTION generated by turbulent flow along a rough wall is higher than that of a smooth wall. As a proof, consider two boundary layers that exhibit identical mean velocities $\overline{u}^{\triangleleft}$ at a given location $z^{\triangleleft}$ within the logarithmic overlap region:

$$\overline{u}^{s}\left(z = z^{\triangleleft}\right) = \overline{u}^{r}\left(z = z^{\triangleleft}\right) \quad .$$

The superscripts s and r denote smooth and rough walls, respectively. Redimensionalizing the logarithmic overlap law yields

$$\begin{aligned}
u_\tau^{s}\left(\frac{1}{\kappa}\ln\frac{z^{\triangleleft}u_\tau^{s}}{\nu} + 5.1\right) &= u_\tau^{r}\left(\frac{1}{\kappa}\ln\frac{z^{\triangleleft}u_\tau^{r}}{\nu} + 5.1 - \Delta u^+\right) \\
\underbrace{u_\tau^{s}\left(\frac{1}{\kappa}\ln\frac{z^{\triangleleft}u_\tau^{s}}{\nu} + 5.1\right)}_{f\left(u_\tau^{s}\right)} &= \underbrace{u_\tau^{r}\left(\frac{1}{\kappa}\ln\frac{z^{\triangleleft}u_\tau^{r}}{\nu} + 5.1\right)}_{f\left(u_\tau^{r}\right)} - u_\tau^{r}\cdot\Delta u^+ \\
f\left(u_\tau^{r}\right) - f\left(u_\tau^{s}\right) &= u_\tau^{r}\cdot\Delta u^+
\end{aligned}$$

Since f is monotonically increasing with u_τ and the right-hand side $u_\tau^{r} \cdot \Delta u^+$ can only take positive values, it follows that

$$u_\tau^{r} > u_\tau^{s} \quad .$$

Reexamining the definition of the friction velocity $u_\tau = \sqrt{\tau_w/\varrho}$ thus confirms that the wall shear stress at the rough wall is higher than that of the smooth wall. Due to the analogy between the velocity and thermal boundary layers, the roughness will consequently increase the convective heat transfer. Correctly capturing the roughness effect is therefore a paramount requirement for simulating ice accretion. Figure 10.8 underlines this fact. Several icing simulations are conducted with TAUICE assuming different magnitudes of the ice roughness.[16] The influence on the ice shape is obvious.

[16] Jan Steiner and Stephan E. Bansmer. Ice Roughness and its Impact on the Ice Accretion Process. In *8th AIAA Atmospheric and Space Environments Conference*, page 3591, 2016

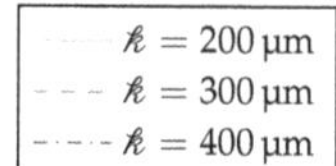

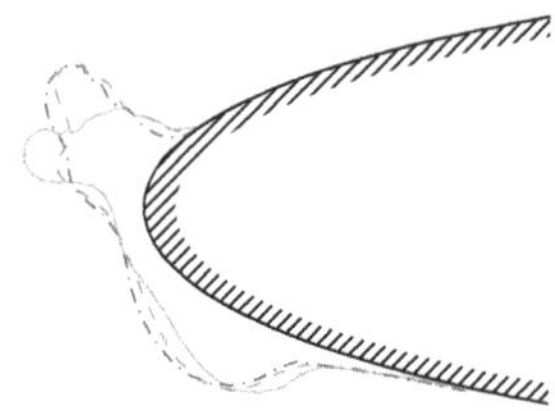

Figure 10.8: Ice accretion simulation with TAUICE for an NACA0012 aerofoil, test case #I1, with different assumptions for the magnitude of ice roughness. Data from STEINER and BANSMER (2016).

10.6 *Roughness and entropy*

One water droplet with a diameter of 20 µm is composed of approximately 1.4×10^{14} water molecules. For typical icing conditions, a number of

$$N_d = \frac{\mathrm{LWC}\, U_\infty A \Delta t}{\varrho_d \frac{\pi}{6} \mathrm{MVD}^3} = \frac{1\,\mathrm{gm^{-3}} \cdot 100\,\mathrm{m\,s^{-1}} \cdot 2\,\mathrm{m^2} \cdot 1\,\mathrm{s}}{1000\,\mathrm{kg\,m^{-3}} \cdot \frac{\pi}{6} \cdot (20\,\mu\mathrm{m})^3} = 4.8 \times 10^{10}$$

droplets are impinging on an aircraft wing each second. Despite this exceedingly high number of microscopic entities and randomness, the macroscopic ice accretion will be the same for identical boundary conditions. Roughness formation seems to be one of the interlinks between microscopic fluctuations and macroscopic ice growth. On the one hand, it incorporates randomness. Even for repetitive icing wind tunnel experiments, the individual roughness elements will never be located at the same place for each icing run. On the other hand, the roughness patterns resemble each other, and a certain degree of order becomes visible. Indeed, *emergence* over a broad range of scales is evident; complex spatial patterns emanate from interacting microstructures.

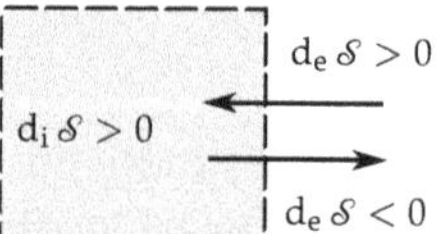

Figure 10.9: Entropy balance of an open, irreversible system.

ENTROPY is one of the thermodynamic state functions and provides a framework for exploring emergence. CLAUSIUS first introduced the concept of entropy – deduced from the Greek word for transformation – while analysing Carnot cycles. For this *reversible* process, he found the relation $d\,\mathcal{S} = d\,Q/T$; an infinitesimal change in entropy $d\,\mathcal{S}$ is caused by an infinitesimal heat flux $d\,Q$ at temperature T. With a modern formalism[17] that involves a control volume approach, the theory can be extended to cover *irreversible* processes of open systems; see Figure 10.9. The entropy change is hence composed of two parts:

$$d\,\mathcal{S} = d_i\,\mathcal{S} + d_e\,\mathcal{S} \quad ,$$

where $d_e\,\mathcal{S}$ characterizes the entropy change caused by an exchange of energy and matter over the boundary and $d_i\,\mathcal{S}$ denotes the entropy produced by an irreversible process inside the control volume. The

[17] Ilya Prigogine. *Introduction to thermodynamics of irreversible processes*. Interscience Publ., 3. ed. edition, 1967. ISBN 0470699280

second law of thermodynamics then requires

$$d_i \mathcal{S} > 0 \quad . \tag{10.2}$$

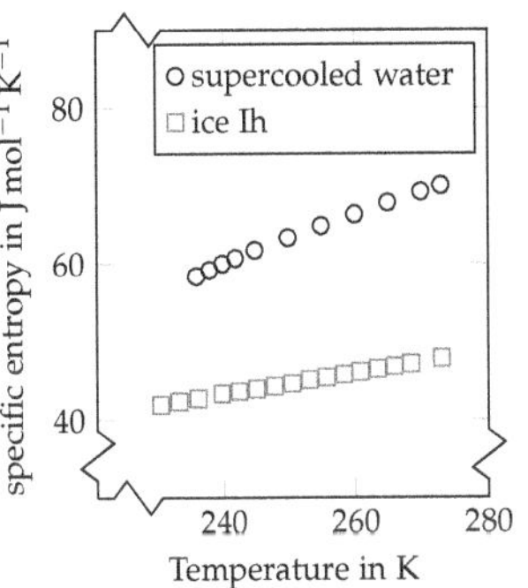

Figure 10.10: Specific entropy of supercooled water and ice. Data from KOSTINSKI and CANTRELL (2008).

Note that the foundation of this entropy concept is entirely macroscopic. It is therefore possible to estimate the entropy of supercooled water and ice by measuring temperatures and heat fluxes; see Figure 10.10. Apparently, the entropy of supercooled water is larger than that of ice; due to the release and export of latent heat, the entropy of water decreases while freezing: $d\mathcal{S} < 0$. Appreciating solidification as an irreversible process – supercooled water can turn into ice but not vice versa – equation (10.2) requires $d_i \mathcal{S} > 0$. In conclusion, the entropy difference between supercooled water and ice must be exported over the boundary, i.e., $d_e \mathcal{S} < 0$.

ENTROPY can also be considered from a microscopic perspective. Founded on statistical mechanics, BOLTZMANN[18] postulated for an isolated system in thermal equilibrium the famous relation between entropy and a number of accessible microstates W, given by

$$\mathcal{S} = k_B \ln W \quad ,$$

where $k_B = 1.3807 \times 10^{-23}\,\mathrm{J\,K^{-1}}$ is called the Boltzmann constant. When contemplating the number of microstates as microscopic degrees of freedom, structures with a lower entropy are more constrained, thus implying an increased order. Consequently, the lower entropy of ice suggests that the molecular structure of ice is more ordered than that of supercooled water.

[18] Ludwig Boltzmann. Weitere Studien über das Wärmegleichgewicht unter Gasmolekülen. *Sitzungsber. Akad. Wiss. (Wien)*, 66(11):275–370, 1872

THE TERM *order* has to be used with great care.[19] Imagine a stack of matches. All matches are translationally and orientationally disordered. When shaking the stack, the matches will reorient themselves to form a parallel arrangement; they become orientationally ordered. The orientational entropy thus decreases. However, when considering the total volume of the stack before shaking, the available space of any one match has drastically increased after shaking, hence providing more translational freedom. Indeed, the translational entropy increases in magnitude, exceeding the loss of orientational entropy.[20] Although the matches seem ordered from a macroscopic perspective, the entropy is counterintuitively increased.

[19] Daan Frenkel. Entropy-driven phase transitions. *Physica A: Statistical Mechanics and its Applications*, 263(1):26–38, 1999. DOI: https://doi.org/10.1016/S0378-4371(98)00501-9. Proceedings of the 20th IUPAP International Conference on Statistical Physics

[20] Lars Onsager. The effects of shape on the interaction of colloidal particles. *Annals of the New York Academy of Sciences*, 51(4):627–659, 1949. DOI: 10.1111/j.1749-6632.1949.tb27296.x

BY COMBINING approaches from information theory[21] and statistical mechanics, a principle of *maximum entropy production* was hypothesized for both reversible[22] and irreversible processes. Figuratively, this principle suggests macroscopic order through microscopic disorder.

[21] Claude E. Shannon. A mathematical theory of communication. *The Bell System Technical Journal*, 27(3): 379–423, 1948. DOI: 10.1002/j.1538-7305.1948.tb01338.x

[22] Edwin T. Jaynes. Information Theory and Statistical Mechanics. *Physical Review*, 106:620–630, 1957. DOI: 10.1103/PhysRev.106.620

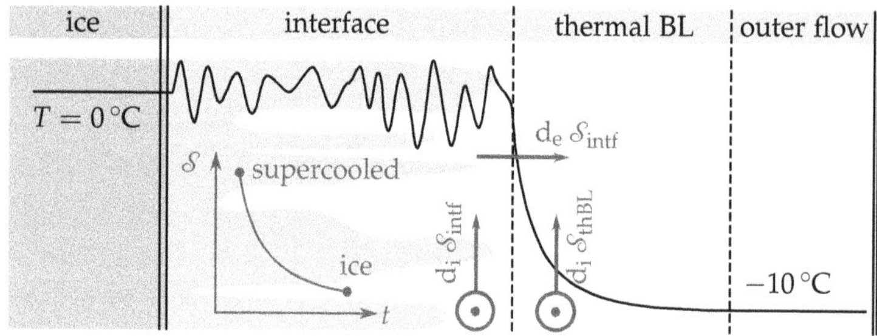

Figure 10.11: Entropy production and transport in the vicinity of the ice surface.

SUMMARIZING THE ABOVE IDEAS for the formation of ice roughness yields the illustration in Figure 10.11. The airflow and accreted ice are separated by an interface that includes roughness elements. Inside the interface region, solidification takes place, thus decreasing the entropy. Small temperature fluctuations around $T_\mathrm{m} = 0\,°\mathrm{C}$ due to dendritic growth cause $\mathrm{d_i}\,\mathcal{S}_\mathrm{intf} > 0$. The release of entropy is exported into the thermal boundary layer of the airflow. Heat conduction into the existing ice layer is neglected. The balance between the exported and produced entropy can be estimated[23] by $\mathrm{d_i}\,\mathcal{S}_\mathrm{intf} \lessapprox 0.1 \cdot |\mathrm{d_e}\,\mathcal{S}_\mathrm{intf}|$. The temperature gradient in the thermal boundary layer causes further production of entropy $\mathrm{d_i}\,\mathcal{S}_\mathrm{thBL} > 0$. Although not fully proven, recent studies[24] suggest entropy-driven roughness morphologies through maximizing $\mathrm{d_i}\,\mathcal{S}$. In particular, the tips of individual roughness elements seem to favour entropy production.[25] Confirmation of the hypothesis of a maximum entropy production principle could significantly improve the computational prediction of ice accretion.

[23] Alexander Kostinski and Will Cantrell. Entropic aspects of supercooled droplet freezing. *Journal of the Atmospheric Sciences*, 65(9):2961–2971, 2008

[24] Adrian Hill. Entropy production as the selection rule between different growth morphologies. *Nature*, 348: 426–428, 1990

[25] Heinz Herwig, Daniel Gloss, and Tammo Wenterodt. Flow in Channels With Rough Walls – Old and New Concepts. *Heat Transfer Engineering*, 31(8):658–665, 2010. DOI: 10.1080/01457630903466605

10.7 *Heat and mass balance according to the Messinger Model*

Having discussed the initial phase of ice accretion on a phenomenological basis, it is very interesting to compare the observations with the computational results of MESSINGER's thermodynamic balance. The analysis draws on test case #I2, an NACA0012 aerofoil at a 3.5° angle of attack that is initially clean.

THE INDIVIDUAL MASS CONTRIBUTIONS of the incipient icing process are depicted in the lower part of Figure 10.12. Note that the ordinate is plotted on a logarithmic scale, which ensures the visibility of all mass terms. The slope of the impinging mass flow $\dot{m}_\mathrm{imp}$ is already familiar due to the proportionality to the collection efficiency β, which is discussed in Section 10.2. The maximum is slightly above the stagnation point because this area of the aerofoil directly faces the trajectories of the impinging droplets.

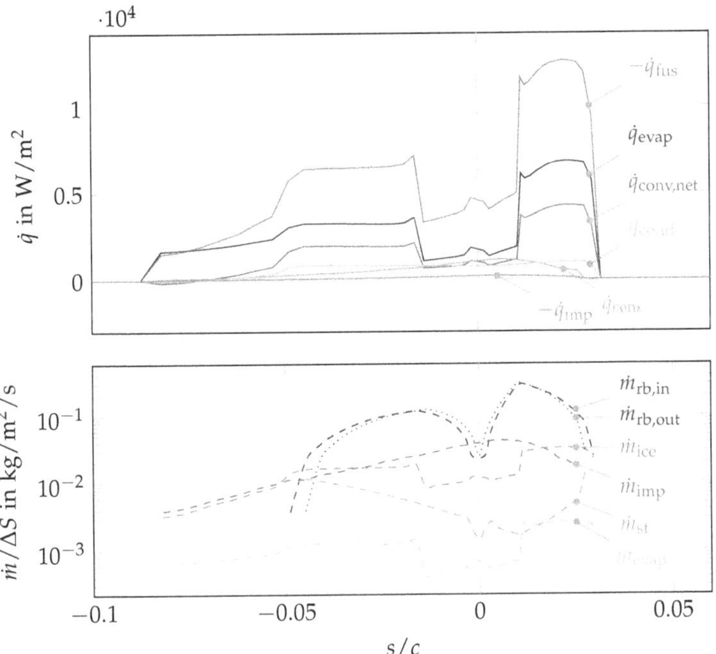

Figure 10.12: Heat and mass contribution according to the Messinger model for an NACA0012 aerofoil, test case #I1. Data provided by JAN STEINER.

THE AMOUNT OF FREEZING WATER is indicated by $\dot{m}_{ice}$. It has a local minimum in the vicinity of the stagnation point $s = 0$ and rises abruptly at $s/c = -0.015$ and $s/c = 0.01$. This is where the glaze ice horns will later form. Liquid water – assumed to exist as a thin film above the ice layer – is called runback, denoted by $\dot{m}_{rb,in}$ and $\dot{m}_{rb,out}$. Near the stagnation point, the amount of runback is nearly one order of magnitude larger than the freezing water mass $\dot{m}_{ice}$. Aerodynamic shear stress at the liquid/air interface drives the runback downstream. Usually, the film velocity $\bar{u}_f$ is significantly smaller than the free-stream velocity U_∞. A shallow water model[26] predicts $\bar{u}_f/U_\infty < 1\,\%$. At $s/c = -0.05$, the runback suddenly decreases and vanishes further downstream. From this location on, nearly all water freezes upon impact, and only slight mass losses due to evaporation occur.

[26] Yves Bourgault, Héloïse Beaugendre, and Wagdi G. Habashi. Development of a shallow-water icing model in FENSAP-ICE. *Journal of Aircraft*, 37(4): 640–646, 2000

ANALYSING THE HEAT BALANCE provides further insights into the MESSINGER model; see the upper part of Figure 10.12. The release of latent heat $\dot{q}_{fus}$ during the solidification of liquid water is the most dominant part of the heat balance. By comparison, the kinetic energy of impacting droplets $\dot{q}_{imp}$, which also accounts for an increase in heat, is two orders of magnitude lower.

PRIMARILY, EVAPORATION AND CONVECTION compensate for the heat increase. Their slopes correlate with each other due to the similarity between the thermal and concentration boundary layers. Recalling Section 9.4, the heat-transfer coefficient $\hbar$ is the common factor in both heat contributions, i.e.,

$$\dot{q}_{\mathrm{conv,net}} = \hbar \cdot \mathfrak{A}_{\mathrm{bead}} \cdot (T_S - T_{\mathrm{rec}}) \quad ,$$
$$\dot{q}_{\mathrm{evap}} = \dot{m}_{\mathrm{evap}}(\hbar) \cdot L_{\mathrm{evap}} \cdot \frac{1}{\Delta S} \quad .$$

Because of the high amount of evaporative cooling, the need to capture the runback dynamics becomes evident. Furthermore, both convective and evaporative heat transfer are closely linked with the mass of freezing water $\dot{m}_{\mathrm{ice}}$. In particular, the sudden increase in heat transfer at $s/c = -0.015$ and $s/c = 0.01$ directly corresponds with the jump in freezing water. Therefore, modelling of the heat transfer coefficient is of paramount importance to correctly capture the size and location of glaze ice horns.

c_f	Skin-friction coefficient
$c_{p,\mathrm{air}}$	Constant-pressure Specific heat of air
$\mathrm{Pr}_{\mathrm{air}}$	Prandtl number of air
Pr_T	Turbulent Prandtl number
Re_k	Roughness Reynolds number
$U_{\mathrm{e,air}}$	Boundary-layer edge velocity
U_{k}	Boundary layer velocity at roughness height k
δ_{thBL}	Thermal boundary-layer thickness
k_{air}	Thermal conductivity of air

THE HEAT TRANSFER MODEL in TAUICE is based on several empirical correlations that are adapted from NASA's icing code LEWICE. A distinction between laminar and turbulent flow is made, yielding

$$\hbar_{\mathrm{lam}} = \frac{2k_{\mathrm{air}}}{\delta_{\mathrm{thBL}}}$$
$$\hbar_{\mathrm{turb}} = \frac{\frac{c_f}{2}\varrho_{\mathrm{air}}\, U_{\mathrm{e,air}}\, c_{\mathrm{p,air}}}{\mathrm{Pr}_T + 0.52\sqrt{\frac{c_f}{2}}\,\mathrm{Re}_k^{0.45}\,\mathrm{Pr}_{\mathrm{air}}^{0.8}} \quad .$$

Switching between the laminar and turbulent model is performed by a modified transition criterion of BRASLOW and VON DOENHOFF[27] that incorporates surface roughness. It compares

$$\mathrm{Re} = \frac{\varrho_{\mathrm{air}}\, k\, U_{\mathrm{k}}}{\mu_{\mathrm{air}}} \Longleftrightarrow \mathrm{Re}_{\mathrm{crit}} \quad , \tag{10.3}$$

[27] Albert E. Von Doenhoff and Albert L. Braslow. The effect of distributed surface roughness on laminar flow. In *Boundary layer and flow control*, pages 657–681. Elsevier, 1961

where the critical Reynolds number $\mathrm{Re}_{\mathrm{crit}}$ is defined by

$$\mathrm{Re}_{\mathrm{crit}} = \begin{cases} 3834.2 - 1.9864 \times 10^5 \left|\frac{s}{c}\right| + \cdots & \\ \cdots + 3.2812 \times 10^6 \left|\frac{s}{c}\right|^2 - 6.9994 \times 10^6 \left|\frac{s}{c}\right|^3 & \left|\frac{s}{c}\right| < 0.035 \\ 600 & \left|\frac{s}{c}\right| \geq 0.035 \end{cases} \quad .$$

These Reynolds numbers together with the heat transfer coefficient $\hbar$ are plotted for testcase #I2 in Figure 10.13. At the transition locations, i.e., for $\mathrm{Re} = \mathrm{Re}_{\mathrm{crit}}$, the heat transfer coefficient suddenly increases, and so does the mass of freezing water. In the present model, the laminar/turbulent transition thus has a major influence on the location of the glaze ice horns.

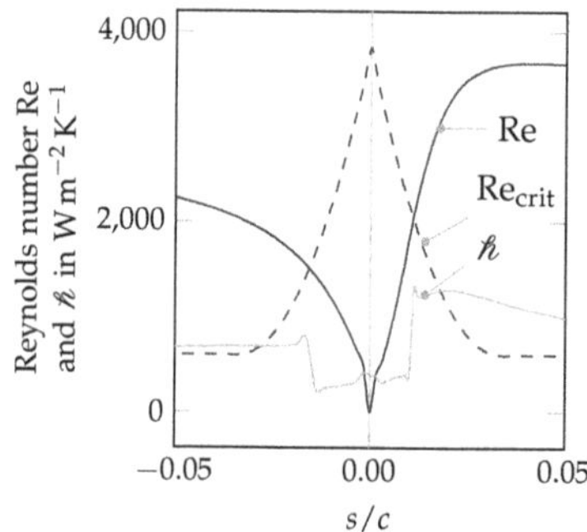

Figure 10.13: Transition evaluation in TAUICE by comparing Re and $\mathrm{Re}_{\mathrm{crit}}$. At the transition points, the heat transfer coefficient suddenly increases. Test case #I1. Data provided by JAN STEINER.

10.8 *Transition is not the driving force*

Revisiting Figure 10.4, a zonal distribution of roughness is observed during glaze ice accretion. The smooth region close to the stagnation line transitions abruptly into a zone of large roughness elements – *beads*. Is this discontinuity in the roughness generated by a jump in heat transfer caused by the laminar-turbulent transition? To answer this question, BRAGG et al. applied artificial hemispherical roughness, similar to that of the initial glaze ice accretion, on an NACA0012 aerofoil.[28] The developing boundary layer flow was studied using hot-wire anemometry in a dry wind-tunnel test. Astonishingly, it was found that the roughness elements trigger transitional flow, but no fully developed turbulent boundary layer existed at the roughness element location. Consequently, a jump condition between laminar/turbulent flow for modelling the heat transfer coefficient, as suggested by equation (10.3), does not capture the physics of incipient ice accretion.

[28] Michael Bragg, Matthew Cummings, and Chad Henze. Boundary-layer and heat-transfer measurements on an airfoil with simulated ice roughness. In *34th Aerospace Sciences Meeting and Exhibit*, number AIAA-96-0866 in Aerospace Sciences Meetings. American Institute of Aeronautics and Astronautics, August 1996. DOI: 10.2514/6.1996-866

WETTING has to be reconsidered to explain the roughness formation. Surface tension is one of the governing parameters of wetting. BILANIN and ANDERSON conducted icing wind tunnel experiments[29] in which surfactants were added to the spray water to reduce the surface tension. Significant changes are observed in the resulting ice shapes. With decreasing surface tension, the glaze ice horns shift towards the leading edge. Researchers thus became aware of including the dynamics of droplet impact and wetting in their current state-of-the art models of ice roughness.

[29] Alan Bilanin and David Anderson. Ice accretion with varying surface tension. In *33rd Aerospace Sciences Meeting and Exhibit Read More: https://arc.aiaa.org/doi/10.2514/6.1995-538*, number AIAA-95-0538 in Aerospace Sciences Meetings. American Institute of Aeronautics and Astronautics, August 1995. DOI: 10.2514/6.1995-538

10.9 *Roughness models*

Despite the random distribution of individual roughness elements, a statistical analysis shows recurring properties when successively repeating an icing experiment with identical boundary conditions. Therefore, roughness models can be deduced: for instance, a mean roughness height $\bar{k}$ as a function of time t_{acc} and position s.

THE NECESSITY of such models was increasingly recognized in the 1980s, when a second generation of icing codes was in preparation.[30] A large number of systematic icing wind tunnel experiments were conducted to derive empirical roughness correlations. The model of SHIN and BOND[31] is the most widely used approximation. It provides a single average representative sand-grain roughness value as a function of the MVD, LWC, temperature T_∞ and velocity U_∞.

[30] Lasse Makkonen. Heat transfer and icing of a rough cylinder. *Cold Regions Science and Technology*, 10(2): 105–116, 1985; and Pascal Personne and Christophe Duroure. Influence de la rugosité du givre sur sa croissance par la captation des goettes d'eau surfondue: application au givrage des cable électriques. *Le Journal de Physique Colloques*, 48(C1):C1–389, 1987

[31] Jaiwon Shin and Thomas H. Bond. Experimental and computational ice shapes and resulting drag increase for a NACA 0012 airfoil. Technical Report 19930018251, NASA Lewis Research Center, 1992

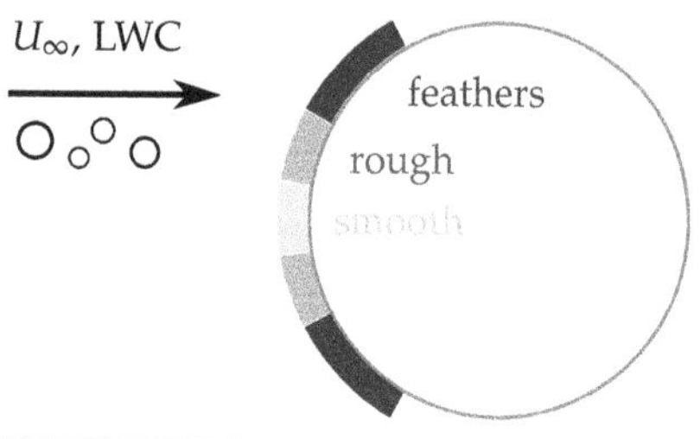

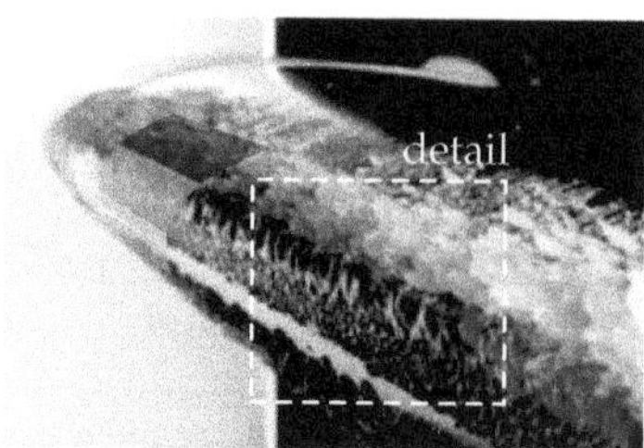

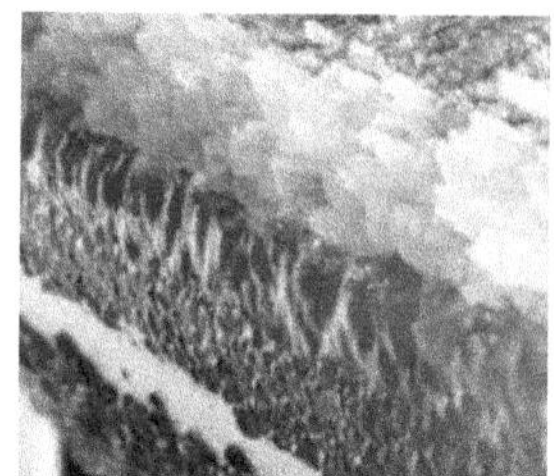

Figure 10.14: Different regions of ice roughness. *Left:* Model according to ANDERSON and Shin (1997). *Right:* Photography of roughness for test case #I4 by JUAN VELANDIA.

SEVERAL EXPERIMENTS[32,33], however, revealed that the distribution of roughness for glaze ice conditions varies along the aerofoil contour and over time, and hence,

$$\left.\begin{array}{r} k_{\text{bead}} \\ d_{\text{bead}} \end{array}\right\} = \vec{f}\left(A_c, f_0, s\right) \quad ,$$

where k_{bead} and d_{bead} represent the bead height and diameter. $A_c = A_c\left(t_{\text{acc}}\right)$ denotes the accumulation parameter, and f_0 is the stagnation line freezing fraction; both parameters are defined in Section 2.2. Figure 10.14 illustrates the derived roughness model. Close to the stagnation point, the ice surface is very smooth, and after a region of transition (not to be confused with the laminar-turbulent flow transition), large roughness elements are visible. Moreover, the significance of the water film above and around the roughness elements was recognized.[34] Water films promote evaporation, which represents a significant heat flux in the thermodynamic balance of ice accretion, as already demonstrated in Figure 10.12. Consequently, new roughness models were developed that address the physical interplay between the water film and roughness elements. Beyond the triple-deck theory,[35] the newest models focus on bead formation and dynamics.[36] Driven by microphysical considerations, these models consider surface tension effects, merging of neighbouring beads, rivulets and continuous water films. Capturing the accompanying solidification process on this microscopic scale is still a challenge for the models.

SUPERCOOLED LARGE DROPLETS (SLD) with an MVD larger than 50 µm pose a new challenge to aircraft certification. On the one hand, their higher inertia shifts the impingement limits further downstream. On the other hand, additional physical phenomena, e.g., the droplet impact at high Weber numbers, must be considered to simulate ice accretion; see also the discussions in part II. Consequently, bead formation is affected. Figure 10.15 shows a preliminary roughness map composed of several SLD icing experiments versus roughness data for lower MVDs. A general trend of increasing roughness

[32] William Olsen and Ernie Walker. Experimental evidence for modifying the current physical model for ice accretion on aircraft surfaces. Technical Report NASA-TM-87184, NASA Lewis Research Center, 1986

[33] R. John Hansman and Stephen R. Turnock. Investigation of surface water behavior during glaze ice accretion. *Journal of Aircraft*, 26(2):140–147, 1989

[34] Tim G. Myers and Chris P. Thompson. Modeling the flow of water on aircraft in icing conditions. *AIAA Journal*, 36(6): 1010–1013, 1998

[35] Jen-Ching Tsao and Alric P. Rothmayer. Application of triple-deck theory to the prediction of glaze ice roughness formation on an airfoil leading edge. *Computers & Fluids*, 31(8):977–1014, 2002

[36] Guy Fortin, Adrian Ilinca, Jean-Louis Laforte, and Vincenzo Brandi. New Roughness Computation Method and Geometric Accretion Model for Airfoil Icing. *Journal of Aircraft*, 41(1):119–127, January 2004. DOI: 10.2514/1.173; and Giulio Croce, Erika De Candido, Wagdi G. Habashi, Jeffrey Munzar, Martin S. Aubé, Guido S. Baruzzi, and Cristhian N. Aliaga. FENSAP-ICE: Analytical Model for Spatial and Temporal Evolution of In-Flight Icing Roughness. *Journal of Aircraft*, 47(4):1283–1289, July 2010. DOI: 10.2514/1.47143

with increasing accumulation parameter is visible. Compared to experiments with low MVD, the roughness height for SLD conditions reaches an asymptote at larger accumulation parameters. Further studies are currently underway to incorporate SLD-roughness models into industrial icing codes.

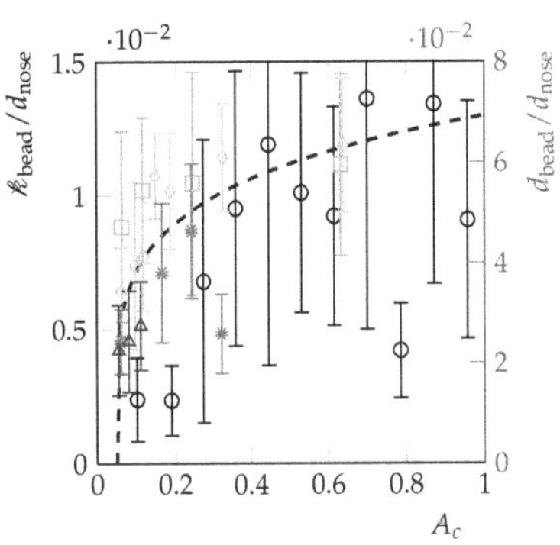

Figure 10.15: Bead height k_{bead} and diameter d_{bead} over accumulation parameter for SLD icing cases #I5 (black), STEINER and BANSMER (2016); and FAR/CS25 Appendix C icing cases with lower MVD (blue), ANDERSON et al. (1998)

10.10 Glaze ice horns

In Section 10.7, the heat and mass balance according to the MESSINGER model were evaluated for the initial stage of ice accretion using TAUICE. Despite the shortcomings of the applied methodology for estimating heat transfer, the simulated growth of the glaze ice horns is adequately captured and matches NASA's experimental results.[37] This is because the underlying heat transfer model adopted from the LEWICE approach is calibrated using a broad range of experimental results from NASA's icing research wind tunnel.

[37] William Wright. Validation Results for LEWICE 3.0. Aerospace Sciences Meetings. American Institute of Aeronautics and Astronautics, January 2005. DOI: 10.2514/6.2005-1243

SINCE TAUICE successively repeats the execution of the flow, trajectory and thermodynamic solvers, the heat and mass balance of ice accretion can be monitored over time. Selected results for the accumulation times $t_{acc} = \{0, 120\,s, 240\,s, 360\,s\}$ are plotted in Figure 10.16. On the left, the mass flow of impinging droplets is highlighted, which is proportional to the collection efficiency β. For $t_{acc} = 0$, the clean aerofoil exhibits one single maximum of β close to the stagnation point. Advancing in time, two maxima emerge representing the glaze ice horns. Their elevated shape in turn promotes the collection of water. At the same time, they prevent droplet impact further downstream. The impingement limits, initially located at $s/c = -0.10$ and $s/c = 0.025$, shift upstream.

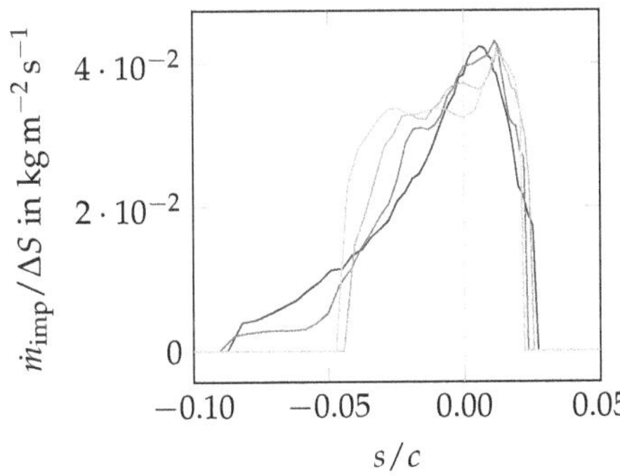

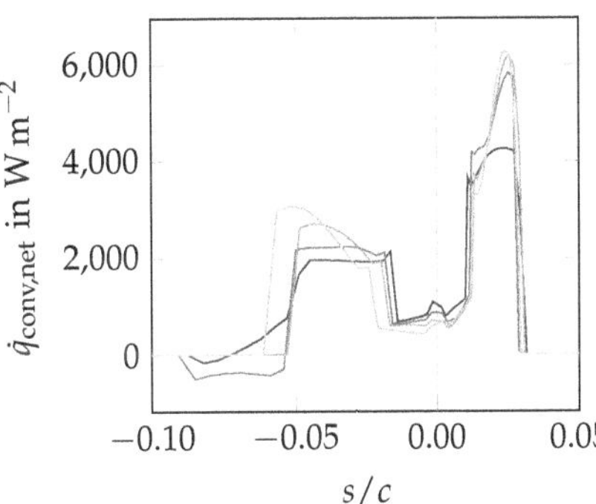

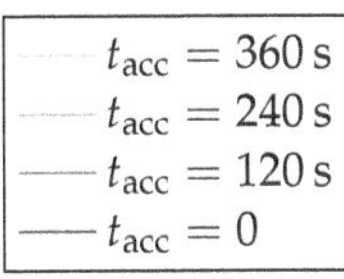

Figure 10.16: Messinger balance of ice accretion over time for test case #I1. *Left:* Mass flow rate of impinging droplets. *Right:* Convective heat transfer along the aerofoil. Data provided by JAN STEINER.

CONVECTIVE HEAT TRANSFER, as shown in the right part of Figure 10.16, is strongly bound to the transition criterion of equation (10.3), thus enforcing the growth direction of the glaze ice horns. Towards the trailing edge of the horns, an increase in heat transfer is observed. This increase is caused by the airflow, which suddenly increases in velocity after clearing the vicinity of the ice accretion, as shown by the sketched streamlines in Figure 10.17.

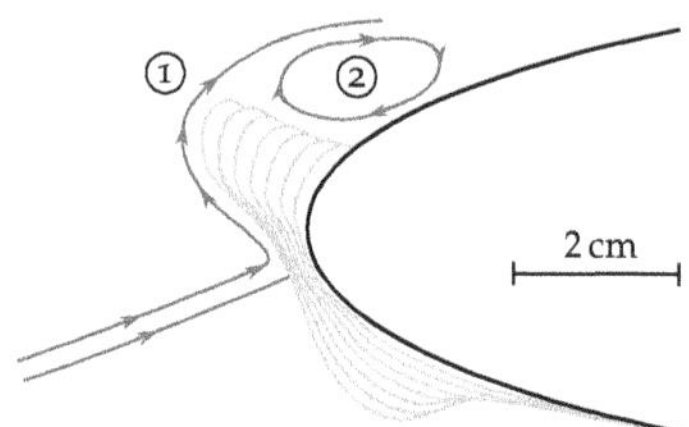

Figure 10.17: Growth of ice horns, simulated by TAUICE, test case #I1. The very high velocities at (1) promote increased convective heat transfer at the rear part of the horns. Further downstream (2), the flow separates. Data provided by JAN STEINER.

SHORTLY AFTER, the flow separates because the glaze ice horns act in a manner similar to a backward-facing step. The aerodynamic performance deteriorates with increasing accumulation time t_{acc}; see Figure 10.18. The stalled flow decreases the lift, and at the same time, the pressure drag increases. Within seven minutes, the lift-to-drag ratio is reduced by 50 %. Obviously, glaze ice accretion is considered a severe threat to the safety of flight.

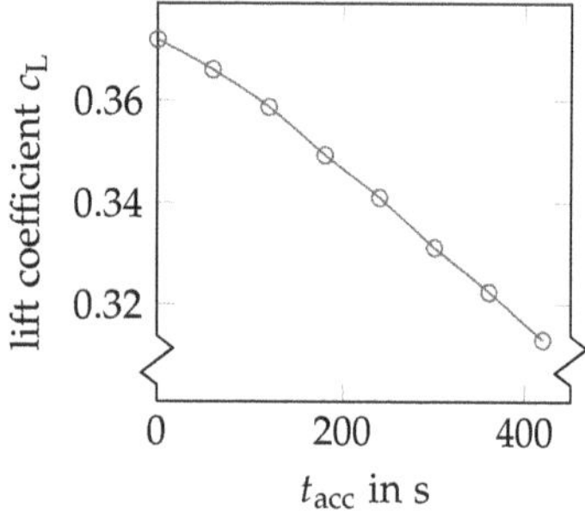

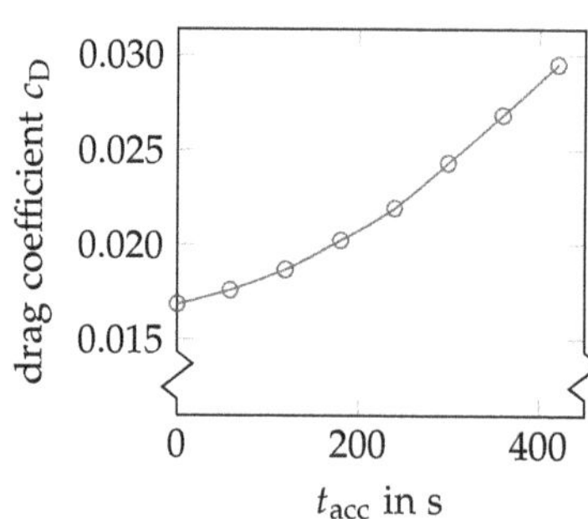

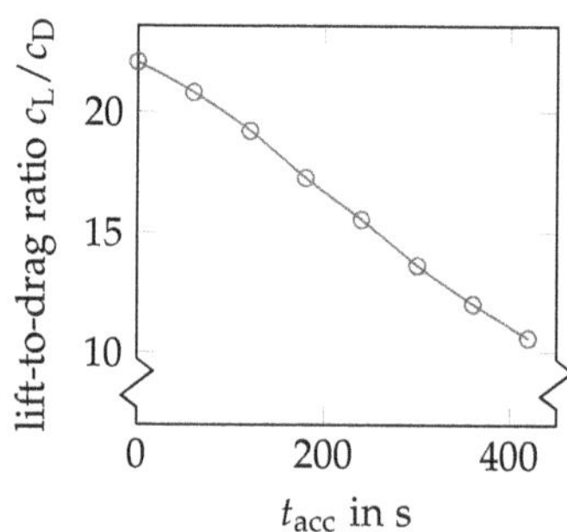

Figure 10.18: The growth of glaze ice horns deteriorates the aerodynamic performance of the NACA 0012 aerofoil. TAUICE computation of test case #I1, data by JAN STEINER.

Summary

Ice accretion is a process that can occur on aerodynamic surfaces as the aircraft flies through clouds of supercooled droplets. In this study, fluid mechanics is used to describe the process of ice accretion from four different perspectives outlined below.

MICROSCOPIC INVESTIGATIONS OF SINGLE IMPACTING DROPLETS. In the first seconds of icing, water droplets collide with dry surfaces at high speed. Splashing produces secondary droplets, which, on the one hand, reduce the deposited mass of the primary droplet and, on the other hand, can collide again with aerodynamic surfaces further downstream. The properties of the secondary droplets are quantified for a wide parameter range. Recently, it has become possible to demonstrate the influence of the surrounding gas phase on the splashing behaviour.[38,39] As the ice accretion of an aerodynamic surface is advancing over time, the droplets collide with a thin water film covering the layer of ice. Droplet impact on a wetted surface generates a much larger splashing, which is caused by the interaction of the inner spreading lamella with the water film on the ice layer. This can be modelled as a propagation of a kinematic discontinuity, which leads to the ejection of an uprising, expanding corona. The secondary droplets of a wet impact are considerably larger compared to a dry impact, and the secondary mass is also larger since the water film originally present on the substrate is contained to a large amount by the secondary droplets. Direct numerical simulations of two-phase flow using the volume-of-fluid (VOF) method for interface tracking complement the present studies.

SYSTEMATIC EXPERIMENTAL INVESTIGATIONS. The Braunschweig Icing Wind Tunnel, which enables the systematic investigation of icing as an overall process, was built as part of this study. Two experimental challenges are elaborated in more detail in the manuscript. The first of these is the characterisation of the cloud of supercooled water droplets in the airflow. For that purpose, several optical measurement techniques were tested in the icing wind tunnel, yielding a

[38] David A. Burzynski and Stephan E. Bansmer. Role of surrounding gas in the outcome of droplet splashing. *Physical Review Fluids*, 4(7):073601, 2019

[39] David A. Burzynski, Ilia V. Roisman, and Stephan E. Bansmer. On the splashing of high-speed drops impacting a dry surface. *Journal of Fluid Mechanics*, 892, 2020. DOI: 10.1017/jfm.2020.168

large data set for probe intercomparison. The second challenge is the measurement of the shape of accreted ice, which is complicated due to the optical transparency for visible light. To this end, an infra-red scanning technique is presented.

Macroscopic observation of ice growth on aerodynamic surfaces. The observation focuses on the formation of glaze ice, which is created at temperatures close to the freezing point of water. This constrains the water to be present in different phase states. The icing takes place in several stages, ranging from initial wetting to the formation of roughness elements and the growth of large-scale ice structures at the leading edge of aerodynamic surfaces. Special attention is given to the formation of ice roughness as it affects convective heat transfer, which is one of the most important factors in the heat balance of ice growth.

Numerical simulations enabling an in-depth analysis of parameters that are difficult to access experimentally and also offer the possibility of predicting ice accretion. A simulation cycle composed of a RANS simulation, a drop trajectory calculation, and a mass and heat balance of the accumulating water is presented in this manuscript. The source of largest uncertainty is due to the modelling of the convective heat transfer, which is influenced by ice roughness and the dynamics of the thin water film above the ice layer. Currently, the modelling is mainly accomplished by empirical and semi-empirical correlations. Due to the exponentially growing computing power, it is to be expected in the coming decades that the underlying physical phenomena of ice accretion will be directly simulated without the use of empirical methods.

Bibliography

Damon G. K. Aboud and Anne-Marie Kietzig. Splashing threshold of oblique droplet impacts on surfaces of various wettability. *Langmuir*, 31(36):10100–10111, 2015.

AC-9C Aircraft Icing Technology Committee. Calibration and Acceptance of Icing Wind Tunnels. Technical report, SAE International, 2015.

Michael Aftosmis, Datta Gaitonde, and Sean Tavares. Behavior of linear reconstruction techniques on unstructured meshes. *AIAA Journal*, 33(11):2038–2049, November 1995. DOI: 10.2514/3.12945.

Heinz-Eberhard Albrecht. Laser Doppler and Phase Doppler Measurement Techniques, 2003.

Steven R. Allmaras, Forrester T. Johnson, and Philippe R. Spalart. Modifications and Clarifications for the Implementation of the Spalart-Allmaras Turbulence Model. In *Seventh International Conference on Computational Fluid Dynamics (ICCFD7), Big Island, Hawaii*, number ICCFD7-1902, July 2012.

Daniel M. Anderson, Geoffrey B. McFadden, and Adam A. Wheeler. Diffuse-interface methods in fluid mechanics. *Annual review of fluid mechanics*, 30(1):139–165, 1998a.

David Anderson and Jaiwon Shin. Characterization of Ice Roughness from Simulated Icing Encounters. In *35th Aerospace Sciences Meeting and Exhibit*, number AIAA-97-0052. American Institute of Aeronautics and Astronautics, 1997. DOI: https://doi.org/10.2514/6.1997-52.

David Anderson, Daniel Hentschel, and Gary Ruff. Measurement and correlation of ice accretion roughness. In *36th AIAA Aerospace Sciences Meeting and exhibit*, page 486, 1998b.

John David Anderson. *Modern compressible flow*. McGraw-Hill series in aeronautical and aerospace engineering. McGraw-Hill, 3rd edition, 2003. ISBN 0071121617.

Charles A. Angell. Supercooled water. *Annual Review of Physical Chemistry*, 34(1):593–630, 1983.

Richard F. W. Bader and Glenyce A. Jones. The electron density distributions in hydride molecules: I. The water molecule. *Canadian Journal of Chemistry*, 41(3):586–606, 1963.

Stephan E. Bansmer, Arne Baumert, Stephan Sattler, Inken Knop, Delphine Leroy, Alfons Schwarzenboeck, Tina Jurkat-Witschas, Christiane Voigt, Hugo Pervier, and Biagio Esposito. Design, Construction and Commissioning of the Braunschweig Icing Wind Tunnel. *Atmospheric Measurement Techniques*, 11:3221–3249, 2018.

Jewel B. Barlow, William H. Rae, and Alan Pope. *Low-speed wind tunnel testing*. Wiley, New York, NY [u.a.], 3. ed edition, 1999. ISBN 0471557749.

Arne Baumert, Stephan E. Bansmer, Pierre Trontin, and Philippe Villedieu. Experimental and numerical investigations on aircraft icing at mixed phase conditions. *International Journal of Heat and Mass Transfer*, 123:957–978, 2018.

Leoplod P. Bayvel and Zdzisław Orzechowski. *Liquid atomization*. Combustion. Taylor & Francis, Washington, 1993. ISBN 0891169598.

Jacob Bear. Introduction to Modeling of Transport Phenomena in Porous Media, 1990.

Héloïse Beaugendre, François Morency, and Wagdi G. Habashi. FENSAP-ICE's Three-Dimensional In-Flight Ice Accretion Module: ICE3D. *Journal of Aircraft*, 40(2):239–247, March 2003. DOI: 10.2514/2.3113.

John W. Biddle, Vincent Holten, Jan V. Sengers, and Mikhail A. Anisimov. Thermal conductivity of supercooled water. *Physical Review E*, 87(4):042302, 2013.

Alan Bilanin and David Anderson. Ice accretion with varying surface tension. In *33rd Aerospace Sciences Meeting and Exhibit Read More: https://arc.aiaa.org/doi/10.2514/6.1995-538*, number AIAA-95-0538 in Aerospace Sciences Meetings. American Institute of Aeronautics and Astronautics, August 1995. DOI: 10.2514/6.1995-538.

David R. Bilodeau, Wagdi G. Habashi, Marco Fossati, and Guido S. Baruzzi. Eulerian modeling of supercooled large droplet splashing and bouncing. *Journal of Aircraft*, 52(5):1611–1624, 2015.

Irmgard Bischofberger, Kelly W. Mauser, and Sidney R. Nagel. Seeing the invisible – Air vortices around a splashing drop. *Physics of Fluids*, 25(9):091110, 2013.

François Blais. Review of 20 years of range sensor development. *Journal of Electronic Imaging*, 13(1):231–243, 2004.

Ludwig Boltzmann. Weitere Studien über das Wärmegleichgewicht unter Gasmolekülen. *Sitzungsber. Akad. Wiss. (Wien)*, 66(11):275–370, 1872.

Thomas H. Bond and David N. Anderson. Manual of Scaling Methods. *NASA/CR-2004-212875*, 2004.

Max Born and Emil Wolf. *Principles of optics*. Cambridge University Press, 7., (expanded) ed., reprinted with corr. edition, 2002. ISBN 9780521642224.

Yves Bourgault, Wagdi G. Habashi, Julien Dompierre, and Guido S. Baruzzi. A finite element method study of Eulerian droplets impingement models. *International Journal for Numerical Methods in Fluids*, 29 (4):429–449, 1999.

Yves Bourgault, Héloïse Beaugendre, and Wagdi G. Habashi. Development of a shallow-water icing model in FENSAP-ICE. *Journal of Aircraft*, 37(4):640–646, 2000.

Jean Boussinesq. *Essai sur la théorie des eaux courantes*. Imprimerie nationale, 1877.

Dean T. Bowden, A. E. Gensemer, and Charles A. Skeen. Engineering summary of airframe icing technical data. Technical report, Federal Aviation Agency, 1964.

Jeremiah U. Brackbill, Douglas B. Kothe, and Charles Zemach. A continuum method for modeling surface tension. *Journal of Computational Physics*, 100(2):335–354, 1992.

Peter Bradshaw, editor. *Turbulence.* Number 12 in Topics in Applied Physics. Springer, Berlin, 2., corr. and updated ed edition, 1978. ISBN 0387088644.

Michael Bragg, Matthew Cummings, and Chad Henze. Boundary-layer and heat-transfer measurements on an airfoil with simulated ice roughness. In *34th Aerospace Sciences Meeting and Exhibit*, number AIAA-96-0866 in Aerospace Sciences Meetings. American Institute of Aeronautics and Astronautics, August 1996. DOI: 10.2514/6.1996-866.

Brian J. Briscoe and Kevin P. Galvin. The effect of surface fog on the transmittance of light. *Solar Energy*, 46 (4):191–197, 1991. DOI: https://doi.org/10.1016/0038-092X(91)90063-3.

Edgar Buckingham. On physically similar systems; illustrations of the use of dimensional equations. *Physical review*, 4(4):345, 1914.

David A. Burzynski and Stephan E. Bansmer. Droplet splashing on thin moving films at high Weber numbers. *International Journal of Multiphase Flow*, 101:202–211, 2018.

David A. Burzynski and Stephan E. Bansmer. Role of surrounding gas in the outcome of droplet splashing. *Physical Review Fluids*, 4(7):073601, 2019.

David A. Burzynski, Ilia V. Roisman, and Stephan E. Bansmer. On the splashing of high-speed drops impacting a dry surface. *Journal of Fluid Mechanics*, 892, 2020. DOI: 10.1017/jfm.2020.168.

Alfonso A. Castrejón-Pita, Rafael Castrejón-Pita, and Ian M. Hutchings. Experimental observation of von Kármán vortices during drop impact. *Physical Review E*, 86(4):045301, 2012.

François Charpin and Guy Fasso. Essais de givrage dans la grande soufflerie de Modane sur maquettes a echelle grandeur et echelle reduite. *L'Aeronautique et l'Astronautique*, (38):23–31, 1972.

Thomas H. Chilton and Allan Philip Colburn. Mass transfer (absorption) coefficients prediction from data on heat transfer and fluid friction. *Industrial & engineering chemistry*, 26(11):1183–1187, 1934.

Chang-Hyun Cho, Jacob Urquidi, and G. Wilse Robinson. Molecular-level description of temperature and pressure effects on the viscosity of water. *The Journal of Chemical Physics*, 111(22):10171–10176, 1999.

Genevieve Comte-Bellot. Hot-wire anemometry. *Annual review of fluid mechanics*, 8(1):209–231, 1976.

Gianpietro E. Cossali, Aldo Coghe, and Marco Marengo. The impact of a single drop on a wetted solid surface. *Experiments in fluids*, 22(6):463–472, 1997.

Giulio Croce, Erika De Candido, Wagdi G. Habashi, Jeffrey Munzar, Martin S. Aubé, Guido S. Baruzzi, and Cristhian N. Aliaga. FENSAP-ICE: Analytical Model for Spatial and Temporal Evolution of In-Flight Icing Roughness. *Journal of Aircraft*, 47(4):1283–1289, July 2010. DOI: 10.2514/1.47143.

Paul I. Crumpton, Pierre Moinier, and Michael B. Giles. An unstructured algorithm for high Reynolds number flows on highly stretched grids. *Numerical methods in laminar and turbulent flow*, pages 561–572, 1997.

Fred E. C. Culick. Comments on a ruptured soap film. *Journal of Applied Physics*, 31(6):1128–1129, 1960.

Germund G. Dahlquist. A special stability problem for linear multistep methods. *BIT Numerical Mathematics*, 3(1):27–43, 1963.

Narsingh Dass and Narendra K. Gilra. Refractive Index of Liquids on Supercooling-Water. *Journal of the Physical Society of Japan*, 21(10):2039–2042, 1966.

Pierre G. de Gennes. Wetting: statics and dynamics. *Reviews of Modern Physics*, 57:827–863, Jul 1985. DOI: 10.1103/RevModPhys.57.827.

Pablo G. Debenedetti. *Metastable liquids*. Physical chemistry. Princeton Univ. Press, 1996. ISBN 0691085951.

Pablo G. Debenedetti. Supercooled and glassy water. *Journal of Physics: Condensed Matter*, 15(45):R1669, 2003.

Christoph Deiler and Thomas Kilian. Dynamic aircraft simulation model covering local icing effects. *CEAS Aeronautical Journal*, Mar 2018. DOI: 10.1007/s13272-018-0291-6.

Suraj S. Deshpande, Lakshman Anumolu, and Mario F. Trujillo. Evaluating the performance of the two-phase flow solver interFoam. *Computational Science & Discovery*, 5(1):014016, 2012.

John R. Dormand and Peter J. Prince. A family of embedded Runge-Kutta formulae. *Journal of Computational and Applied Mathematics*, 6(1):19 – 26, 1980.

Michelle M. Driscoll and Sidney R. Nagel. Ultrafast interference imaging of air in splashing dynamics. *Physical Review Letters*, 107(15):154502, 2011.

Hugh L. Dryden. Review of published data on the effect of roughness on transition from laminar to turbulent flow. *Journal of the Aeronautical Sciences*, 20(7):477–482, 1953.

Benjamin W. Faßmann, Stephan E. Bansmer, Thorsten J. Möller, Rolf Radespiel, and Michael Hartmann. High velocity impingement of single droplets on a dry smooth surface. *Experiments in fluids*, 54(5):1516, 2013.

Alexandre Favre. Equations des gaz turbulents compressibles. *Journal de mecanique*, 4(4):391, 1965.

Rainer Feistel and Wolfgang Wagner. A new equation of state for H_2O ice Ih. *Journal of Physical and Chemical Reference Data*, 35(2):1021–1047, 2006.

Alejandro Feo and Jen-Ching Tsao. The Water Film Weber Number in Glaze Icing Scaling. In *SAE Aircraft & Engine Icing International Conference*, number SAE-2007-01-3295, September 2007.

Karen J. Finstad, Edward P. Lozowski, and Lasse Makkonen. On the median volume diameter approximation for droplet collision efficiency. *Journal of the Atmospheric Sciences*, 45(24):4008–4012, 1988.

Neville Horner Fletcher. The chemical physics of ice, 1970.

Guy Fortin, Adrian Ilinca, Jean-Louis Laforte, and Vincenzo Brandi. New Roughness Computation Method and Geometric Accretion Model for Airfoil Icing. *Journal of Aircraft*, 41(1):119–127, January 2004. DOI: 10.2514/1.173.

Bernard Fox, Michael Boito, John C. Graser, and Obaid Younossi. *Test and evaluation trends and costs for aircraft and guided weapons*. RAND Corporation, 2004. ISBN 0833035401.

Felix Franks. Water, 2000.

Daan Frenkel. Entropy-driven phase transitions. *Physica A: Statistical Mechanics and its Applications*, 263 (1):26–38, 1999. DOI: https://doi.org/10.1016/S0378-4371(98)00501-9. Proceedings of the 20th IUPAP International Conference on Statistical Physics.

Daniela Fritter, Charles M. Knobler, and Daniel A. Beysens. Experiments and simulation of the growth of droplets on a surface (breath figures). *Physical Review A*, 43:2858–2869, Mar 1991. DOI: 10.1103/PhysRevA.43.2858.

Thomas F. Gelder, James P. Lewis, and Stanley L. Koutz. Icing Protection for a Turbojet Transport Airplane: Heating Requirements, Methods of Protection, and Performance Penalties. Technical Report NACA-TN-2866, National Advisory Committee for Aeronautics, 1953.

Roger W. Gent. TRAJICE2: a combined water droplet and ice accretion prediction code for aerofoils. Technical Report RAE Technical Report 90054, Royal Aerospace Establishment, 1994.

Roger W. Gent, Nicholas P. Dart, and James T. Cansdale. Aircraft icing. *Philosophical Transactions of the Royal Society of London A: Mathematical, Physical and Engineering Sciences*, 358(1776):2873–2911, 2000.

Martin E. Glicksman. Free dendritic growth. *Materials Science and Engineering*, 65(1):45–55, 1984.

Sergei Konstantinovich Godunov. A difference method for numerical calculation of discontinuous solutions of the equations of hydrodynamics. *Matematicheskii Sbornik*, 89(3):271–306, 1959.

Xiaoliang Gong and Stephan E. Bansmer. 3-D ice shape measurements using mid-infrared laser scanning. *Optics Express*, 23(4):4908–4926, 2015a.

Xiaoliang Gong and Stephan E. Bansmer. Laser scanning applied for ice shape measurements. *Cold Regions Science and Technology*, 115:64–76, 2015b.

Steven Green. A Study of U.S. Inflight Icing Accidents and Incidents, 1978 to 2002. Aerospace Sciences Meetings. American Institute of Aeronautics and Astronautics, September 2006.

Morton E. Gurtin, Eliot Fried, and Lallit Anand. *The mechanics and thermodynamics of continua*. Cambridge University Press, 2013. ISBN 978-0521405980.

George M. Hale and Marvin R. Querry. Optical constants of water in the 200-nm to 200-μm wavelength region. *Applied optics*, 12(3):555–563, 1973.

John Hallett. The temperature dependence of the viscosity of supercooled water. *Proceedings of the Physical Society*, 82(6):1046, 1963.

R. John Hansman and Stephen R. Turnock. Investigation of surface water behavior during glaze ice accretion. *Journal of Aircraft*, 26(2):140–147, 1989.

Ami Harten. High resolution schemes for hyperbolic conservation laws. *Journal of Computational Physics*, 49 (3):357–393, 1983.

Earl C. Hastings Jr and Leonard M. Weinstein. Preliminary indications of water film distribution and thickness on an airfoil in a water spray. Technical Report NASA-TM-85796, NASA Langley Research Center, 1984.

William M. Haynes. *CRC Handbook of Chemistry and Physics, 93rd Edition*. CRC Press, London, 2016. ISBN 9781439880500.

Thierry Hedde and Didier Guffond. ONERA three-dimensional icing model. *AIAA Journal*, 33(6):1038–1045, June 1995. DOI: 10.2514/3.12795.

Henry Selby Hele-Shaw. Investigation of the nature of surface resistance of water and of stream-line motion under certain experimental conditions. volume XL, page 25. Trans. Inst. Nav. Arch., 1898.

Heinz Herwig, Daniel Gloss, and Tammo Wenterodt. Flow in Channels With Rough Walls – Old and New Concepts. *Heat Transfer Engineering*, 31(8):658–665, 2010. DOI: 10.1080/01457630903466605.

John L. Hess. Panel Methods in Computational Fluid Dynamics. *Annual Review of Fluid Mechanics*, 22(1): 255–274, 1990. DOI: 10.1146/annurev.fl.22.010190.001351.

Peter D. Hicks and Richard Purvis. Air cushioning and bubble entrapment in three-dimensional droplet impacts. *Journal of Fluid Mechanics*, 649:135–163, 2010.

Adrian Hill. Entropy production as the selection rule between different growth morphologies. *Nature*, 348: 426–428, 1990.

Jason P. Hindmarsh, Alexander B. Russell, and Xiao D. Chen. Experimental and numerical analysis of the temperature transition of a suspended freezing water droplet. *International Journal of Heat and Mass Transfer*, 46(7):1199–1213, 2003.

Charles Hirsch. *Numerical computation of internal and external flows*. Elsevier/Butterworth-Heinemann, Amsterdam, 2007.

Cyril W. Hirt and Billy D. Nichols. Volume of fluid (VOF) method for the dynamics of free boundaries. *Journal of Computational Physics*, 39(1):201–225, 1981.

Ian M. Hodge and Charles A. Angell. The relative permittivity of supercooled water. *The Journal of Chemical Physics*, 68(4):1363–1368, 1978.

Raimund Honsek, Wagdi G. Habashi, and Martin S. Aubé. Eulerian modeling of in-flight icing due to supercooled large droplets. *Journal of Aircraft*, 45(4):1290–1296, 2008.

Jan Hrubý, Václav Vinš, Radim Maresš, Jirí Hykl, and Jana Kalová. Surface tension of supercooled water: no inflection point down to - 25°C. *The Journal of Physical Chemistry Letters*, 5(3):425–428, 2014.

Lien-Peng Hsiang and Gerard M. Faeth. Near-limit drop deformation and secondary breakup. *International Journal of Multiphase Flow*, 18(5):635–652, 1992.

Frank P. Incropera and David P. DeWitt. *Fundamentals of heat and mass transfer*. Wiley, New York, 5. ed edition, 2002. ISBN 0471386502.

Raad I. Issa. Solution of the implicitly discretised fluid flow equations by operator-splitting. *Journal of Computational Physics*, 62(1):40–65, 1986.

Georgy P. Ivantsov. Temperature around a spheroidal, cylindrical and acicular crystal growing in a supercooled melt. In *Dokl. Akad. Nauk.*, volume 58, pages 567–569, 1947.

Max Jakob. *Heat transfer*. Wiley, New York, NY, 1949.

Antony Jameson and Eli Turkel. Implicit schemes and LU decompositions. *Mathematics of Computation*, 37 (156):385–397, 1981.

Antony Jameson, Wolfgang Schmidt, and Eli Turkel. Numerical Solution of the Euler Equations by Finite Volume Methods using Runge-Kutta Time-Stepping Schemes. In *AIAA-1981-1259*, 1981.

Hrvoje Jasak. *Error analysis and estimation for finite volume method with applications to fluid flow*. PhD thesis, University of London, 1996.

Edwin T. Jaynes. Information Theory and Statistical Mechanics. *Physical Review*, 106:620–630, 1957. DOI: 10.1103/PhysRev.106.620.

Javier Jiménez. Turbulent flows over rough walls. *Annual Review of Fluid Mechanics*, 36:173–196, 2004.

Christophe Josserand and Sigurdur T. Thoroddsen. Drop impact on a solid surface. *Annual Review of Fluid Mechanics*, 48:365–391, 2016.

Christophe Josserand, Pascal Ray, and Stéphane Zaleski. Droplet impact on a thin liquid film: anatomy of the splash. *Journal of Fluid Mechanics*, 802:775–805, 2016.

Stefan Jung, Manish K. Tiwari, N. Vuong Doan, and Dimos Poulikakos. Mechanism of supercooled droplet freezing on surfaces. *Nature Communications*, 3:615, January 2012.

Pyotr L. Kapitza and Sergey P. Kapitza. Wave flow of thin viscous fluid layers. *Zh. Eksp. Teor. Fiz*, 18(1): 3–28, 1948.

Stefan Kedenburg, Marius Vieweg, Timo Gissibl, and Harald Giessen. Linear refractive index and absorption measurements of nonlinear optical liquids in the visible and near-infrared spectral region. *Optical Materials Express*, 2(11):1588–1611, 2012.

Ki Seong Kim and Sung-Soo Kim. Drop sizing and depth-of-field correction in TV imaging. *Atomization and Sprays*, 4(1), 1994.

John G. Kirkwood and Frank P. Buff. The statistical mechanical theory of surface tension. *The Journal of Chemical Physics*, 17(3):338–343, 1949.

Philip S. Klebanoff. Characteristics of turbulence in a boundary layer with zero pressure gradient. Technical Report NACA Report 1247, NASA–Langley Research Center, Hampton, VA, 1954.

James G. Knudsen and Donald L. Katz. *Fluid dynamics and heat transfer*. McGraw-Hill chemical engineering series. McGraw-Hill, New York [u.a.], 1958.

Alexander Kostinski and Will Cantrell. Entropic aspects of supercooled droplet freezing. *Journal of the Atmospheric Sciences*, 65(9):2961–2971, 2008.

Eric N. Landis and Denis T. Keane. X-ray microtomography. *Materials characterization*, 61(12):1305–1316, 2010.

James S. Langer. Instabilities and pattern formation in crystal growth. *Reviews of Modern Physics*, 52(1):1, 1980.

Irving Langmuir. *The collected works of Irving Langmuir*, volume 10. Pergamon Press, 1961.

Irving Langmuir and Katherine B. Blodgett. *A mathematical investigation of water droplet trajectories*. Number 5418 in Army Air Forces Technical Report. Army Air Forces, Air Materiel Command, Dayton, Ohio, 1946.

Irving Langmuir and Katherine B. Blodgett. *The collected works of Irving Langmuir*, volume 10 - Atmospheric phenomena, chapter Mathematical investigation of water droplet trajectories, pages 335–393. Pergamon press, 1961.

Wendell M. Latimer and Worth H. Rodebush. Polarity and ionization from the standpoint of the Lewis theory of valence. *Journal of the American Chemical Society*, 42(7):1419–1433, 1920.

William M. Leary. "We Freeze to Please": A History of NASA's Icing Research Tunnel and the Quest for Flight Safety. Technical Report NASA/SP-2002-4226, National Aeronautics and Space Administration, 2002.

Nikolai V. Lebedev. *Combating ice on aircraft*. State publishing house of defense industry, People's Commissariat of Aviation Industry, 1939.

Sung P. Lin and Rolf D. Reitz. Drop and spray formation from a liquid jet. *Annual Review of Fluid Mechanics*, 30(1):85–105, 1998.

Harvard Lomax and Joseph L. Steger. Relaxation Methods in Fluid Mechanics. *Annual Review of Fluid Mechanics*, 7(1):63–88, 1975.

Robert MacCormack and Alvin Paullay. Computational efficiency achieved by time splitting of finite difference operators. Aerospace Sciences Meetings. American Institute of Aeronautics and Astronautics, January 1972.

Lasse Makkonen. Heat transfer and icing of a rough cylinder. *Cold Regions Science and Technology*, 10(2): 105–116, 1985.

Lasse Makkonen. Models for the growth of rime, glaze, icicles and wet snow on structures. *Philosophical Transactions of the Royal Society of London A: Mathematical, Physical and Engineering Sciences*, 358(1776):2913–2939, 2000. DOI: 10.1098/rsta.2000.0690.

Tafara E. Makuni, Hardeep S. Kalsi, Paul G. Tucker, and Holger Babinsky. Transitional Shock-Wave/Boundary-Layer Interactions in Intakes at Incidence. In *Advances in Simulation of Wing and Nacelle Stall*, pages 323–333, 2016. DOI: 10.1007/978-3-319-21127-5_19.

Shreyas Mandre and Michael P. Brenner. The mechanism of a splash on a dry solid surface. *Journal of Fluid Mechanics*, 690:148–172, 2012.

Franck S. Marzani, Yvon Voisin, Lew F. C. Lew Yan Voon, and Alain Diou. Calibration of a three-dimensional reconstruction system using a structured light source. *Optical Engineering*, 41(2):484–492, 2002.

Åke Melinder. Properties and other aspects of aqueous solutions used for single phase and ice slurry applications. *International Journal of Refrigeration*, 33(8):1506–1512, 2010.

Bernard L. Messinger. Equilibrium temperature of an unheated icing surface as a function of air speed. *Journal of the Aeronautical Sciences*, 20(1):29–42, 1953.

Giuseppe Mingione and Vincenzo Brandi. Ice Accretion Prediction on Multielement Airfoils. *Journal of Aircraft*, 35(2):240–246, March 1998. DOI: 10.2514/2.2290.

Osamu Mishima and H. Eugene Stanley. The relationship between liquid, supercooled and glassy water. *Nature*, 396(6709):329–335, 1998.

William W. Mullins and Robert F. Sekerka. Morphological stability of a particle growing by diffusion or heat flow. *Journal of Applied Physics*, 34(2):323–329, 1963.

Tim G. Myers and Chris P. Thompson. Modeling the flow of water on aircraft in icing conditions. *AIAA Journal*, 36(6):1010–1013, 1998.

Johann Nikuradse. Strömungsgesetze in rauhen Rohren. *VDI Forschungsheft*, 361, 1933.

William Olsen and Ernie Walker. Experimental evidence for modifying the current physical model for ice accretion on aircraft surfaces. Technical Report NASA-TM-87184, NASA Lewis Research Center, 1986.

Lars Onsager. The effects of shape on the interaction of colloidal particles. *Annals of the New York Academy of Sciences*, 51(4):627–659, 1949. DOI: 10.1111/j.1749-6632.1949.tb27296.x.

David W. Oxtoby. Density functional methods in the statistical mechanics of materials. *Annual Review of Materials Research*, 32(1):39–52, 2002.

Suhas V. Patankar. A calculation procedure for two-dimensional elliptic situations. *Numerical Heat Transfer*, 4(4):409–425, 1981.

Suhas V. Patankar and D. Brian Spalding. A calculation procedure for heat, mass and momentum transfer in three-dimensional parabolic flows. *International Journal of Heat and Mass Transfer*, 15(10):1787–1806, 1972.

Robert H. Perry and Don W. Green, editors. *Perry's chemical engineers' handbook*. McGraw-Hill, New York, 8. ed. edition, 2008. ISBN 9780071422949.

Pascal Personne and Christophe Duroure. Influence de la rugosité du givre sur sa croissance par la captation des goettes d'eau surfondue: application au givrage des cable électriques. *Le Journal de Physique Colloques*, 48(C1):C1–389, 1987.

Victor F. Petrenko and Robert W. Whitworth. *Physics of ice*. Oxford University Press, 2006. ISBN 0198518943.

Marcia K. Politovich. Aircraft Icing Caused by Large Supercooled Droplets. *Journal of Applied Meteorology*, 28(9):856–868, 1989.

Yves Pomeau. *Surface tension: from fundamental principles to applications in liquids and in solids*. PhD thesis, Warsaw School of Statistical Physics, 2013.

Stephen B. Pope. *Turbulent flows*. Cambridge Univ. Press, 12. print. edition, 2015. ISBN 9780521598866.

John A. Pople. The molecular orbital theory of chemical valency. V. The structure of water and similar molecules. In *Proceedings of the Royal Society of London A: Mathematical, Physical and Engineering Sciences*, volume 202, pages 323–336. The Royal Society, 1950.

Martine Prévost and Dominique Gallez. Nonlinear rupture of thin free liquid films. *The Journal of Chemical Physics*, 84(7):4043–4048, 1986.

Ilya Prigogine. *Introduction to thermodynamics of irreversible processes*. Interscience Publ., 3. ed. edition, 1967. ISBN 0470699280.

Hans R. Pruppacher. A new look at homogeneous ice nucleation in supercooled water drops. *Journal of the Atmospheric Sciences*, 52(11):1924–1933, 1995.

George D. Raithby and Gerald E. Schneider. Numerical solution of problems in incompressible fluid flow: treatment of the velocity-pressure coupling. *Numerical Heat Transfer, Part A: Applications*, 2(4):417–440, 1979.

Michael R. Raupach, Robert A. Antonia, and Sundara Rajagopalan. Rough-wall turbulent boundary layers. *Applied Mechanics Reviews*, 44(1):1–25, 1991.

Andrew L. Reehorst and G. Paul Richter. New Methods and Materials for Molding and Casting Ice Formations. *NASA Technical Memorandum*, (100126), 1987.

Osborne Reynolds. On the Extent and Action of the Heating Surface of Stream Boilers. *Proc. Lit. Soc. Manchester*, 14:7–12, 1874.

Osborne Reynolds. On the dynamical theory of incompressible viscous fluids and the determination of the criterion. *Philosophical Transactions of the Royal Society of London. A*, 186:123–164, 1895.

Stephen K. Robinson. Coherent Motions in the Turbulent Boundary Layer. *Annual Review of Fluid Mechanics*, 23(1):601–639, 1991.

Ilia V. Roisman. Inertia dominated drop collisions. II. An analytical solution of the Navier–Stokes equations for a spreading viscous film. *Physics of Fluids*, 21(5):052104, 2009.

Ilia V. Roisman and Cameron Tropea. Impact of a drop onto a wetted wall: description of crown formation and propagation. *Journal of Fluid Mechanics*, 472:373–397, 2002.

Ilia V. Roisman, Kristijan Horvat, and Cameron Tropea. Spray impact: rim transverse instability initiating fingering and splash, and description of a secondary spray. *Physics of Fluids*, 18(10):102104, 2006.

Julius C. Rotta. Das in Wandnähe gültige Geschwindigkeitsgesetz turbulenter Strömungen. *Ingenieur-Archiv*, 18(4):277–280, 1950.

Gary A. Ruff. *Analysis and verification of the icing scaling equations*. PhD thesis, University of Tennessee, Knoxville, 1986.

Henrik Rusche. *Computational fluid dynamics of dispersed two-phase flows at high phase fractions*. PhD thesis, University of London, 2002.

Ruben Scardovelli and Stéphane Zaleski. Direct numerical simulation of free-surface and interfacial flow. *Annual Review of Fluid Mechanics*, 31(1):567–603, 1999.

Walter Schäfer and Cameron Tropea. Time-shift technique for simultaneous measurement of size, velocity, and relative refractive index of transparent droplets or particles in a flow. *Applied Optics*, 53(4):588–597, 2014.

William E. Schiesser. *The Numerical Method of Lines*. Elsevier Science, 2012. ISBN 9780126241303.

Hermann Schlichting. *Boundary-Layer theory*. Springer, Berlin, 9th edition, 2017. ISBN 9783662529171.

Markus Schremb and Cameron Tropea. Solidification of supercooled water in the vicinity of a solid wall. *Physical Review E*, 94:052804, Nov 2016.

Galen B. Schubauer. Turbulent processes as observed in boundary layer and pipe. *Journal of Applied Physics*, 25(2):188–196, 1954.

Dieter Schwamborn, Thomas Gerhold, and Ralf Heinrich. The DLR TAU-code: Recent applications in research and industry. In *ECCOMAS CFD 2006 conference*, 2006.

Richard P. Sear. Nucleation: theory and applications to protein solutions and colloidal suspensions. *Journal of Physics: Condensed Matter*, 19(3):033101, 2007.

Claude E. Shannon. A mathematical theory of communication. *The Bell System Technical Journal*, 27(3): 379–423, 1948. DOI: 10.1002/j.1538-7305.1948.tb01338.x.

Dmitri Sharov, Kazuhiro Nakahashi, Dmitri Sharov, and Kazuhiro Nakahashi. Reordering of 3-D hybrid unstructured grids for vectorized LU-SGS Navier-Stokes computations. In *13th Computational Fluid Dynamics Conference*, page 2102, 1997.

Jaiwon Shin and Thomas H. Bond. Experimental and computational ice shapes and resulting drag increase for a NACA 0012 airfoil. Technical Report 19930018251, NASA Lewis Research Center, 1992.

Roger L. Simpson. Turbulent Boundary-Layer Separation. *Annual Review of Fluid Mechanics*, 21(1):205–232, 1989.

Roger L. Simpson. Junction flows. *Annual Review of Fluid Mechanics*, 33(1):415–443, 2001. DOI: 10.1146/annurev.fluid.33.1.415.

Hannah Sommerwerk, Peter Horst, and Stephan E. Bansmer. Studies on electro impulse de-icing of a leading edge structure in an icing wind tunnel. In *8th AIAA Atmospheric and Space Environments Conference*, page 3441, 2016.

Philippe R. Spalart. Detached-eddy simulation. *Annual Review of Fluid Mechanics*, 41:181–202, 2009.

Philippe R. Spalart and Steven R. Allmaras. A One-Equation Turbulence Model for Aerodynamic Flows. In *AIAA 30th Aerospace Sciences Meeting and Exhibit, Reno, Nevada*, number AIAA-92-0439, January 1992.

James R. Stallabrass. An appraisal of the single rotating cylinder method of liquid water content measurement. Technical report, National Research Council of Canada (NRC), 1978.

Jan Steiner and Stephan E. Bansmer. Ice Roughness and its Impact on the Ice Accretion Process. In *8th AIAA Atmospheric and Space Environments Conference*, page 3591, 2016.

George Gabriel Stokes. Mathematical proof of the identity of the stream lines obtained by means of a viscous film with those of a perfect fluid moving in two dimensions. *Brit. Ass. Rep*, 143, 1898.

Krzysztof Szilder. Simulation of ice accretion on a cylinder due to freezing rain. *Journal of Glaciology*, 40 (136):586–594, 1994. DOI: 10.3189/S0022143000012478.

Gustav Tammann. Über die Grenzen des festen Zustandes IV. *Annalen der Physik*, 307(5):1–31, 1900.

Geoffrey Taylor. The dynamics of thin sheets of fluid. II. Waves on fluid sheets. *Proceedings of the Royal Society of London. Series A, Mathematical and Physical Sciences*, pages 296–312, 1959.

Marie-Jean Thoraval, Kohsei Takehara, Takeharu Goji Etoh, Stéphane Popinet, Pascal Ray, Christophe Josserand, Stéphane Zaleski, and Sigurdur T. Thoroddsen. von Kármán vortex street within an impacting drop. *Physical Review Letters*, 108(26):264506, 2012.

Sigurdur T. Thoroddsen, Takeharu G. Etoh, and Kohsei Takehara. High-speed imaging of drops and bubbles. *Annual Review of Fluid Mechanics*, 40:257–285, 2008.

Albert A. Townsend. The structure of the turbulent boundary layer. In *Mathematical Proceedings of the Cambridge Philosophical Society*, volume 47, pages 375–395. Cambridge University Press, 1951.

Cameron Tropea. Optical particle characterization in flows. *Annual Review of Fluid Mechanics*, 43:399–426, 2011.

Jen-Ching Tsao and Alric P. Rothmayer. Application of triple-deck theory to the prediction of glaze ice roughness formation on an airfoil leading edge. *Computers & Fluids*, 31(8):977–1014, 2002.

Eli Turkel. Improving the Accuracy of Central Difference Schemes. In *11th International Conference on Numerical Methods in Fluid Dynamics*, volume 323 of *Lecture Notes in Physics*, pages 586–591. Springer, 1988.

Hendrik C. van de Hulst. Light Scattering by Small Particles, 2012.

Bram Van Leer. Towards the ultimate conservation difference scheme. II. Monotonicity and conservation combined in a second-order scheme. *Journal of Computational Physics*, 14:361–370, 1974.

Mario Vargas, Jen-Ching Tsao, and Alric Rothmayer. Review of Role of Icing Feathers in Ice Accretion Formation. In *2007 SAE Aircraft and Engine Icing International Conference*. SAE International, sep 2007. DOI: https://doi.org/10.4271/2007-01-3294.

Juan S. Velandia, Alexander Diener, and Stephan E. Bansmer. Porosity of glaze ice from micro computed tomography. To be submitted to Journal of Glaciology.

Christian Vogt, Karim Laihem, and Christopher Wiebusch. Speed of sound in bubble-free ice. *The Journal of the Acoustical Society of America*, 124(6):3613–3618, 2008.

Albert E. Von Doenhoff and Albert L. Braslow. The effect of distributed surface roughness on laminar flow. In *Boundary layer and flow control*, pages 657–681. Elsevier, 1961.

Henry Vu, Darren Banks, and Guillermo Aguilar. Examining viscosity and surface wettability on lamella lift dynamics and droplet splashing. *Atomization and Sprays*, 21(4), 2011.

Bernhard Wagner. Ice Accretion Simulation. Technical Report AGARD report no. 344, North Atlantic Treaty Organization, 1997.

Stephen G. Warren. Optical constants of ice from the ultraviolet to the microwave. *Applied Optics*, 23(8): 1206–1225, 1984.

Frank M. White. *Fluid mechanics*. McGraw-Hill, New York, NY, 8. ed. edition, 2016. ISBN 9780073398273.

David C. Wilcox. *Turbulence Modeling for CFD*, volume 2. DCW Industries, third edition edition, 2006. ISBN 978-1928729082.

William Wright. Validation Results for LEWICE 3.0. Aerospace Sciences Meetings. American Institute of Aeronautics and Astronautics, January 2005. DOI: 10.2514/6.2005-1243.

William Wright and Mark Potapczuk. Semi-Empirical Modelling of SLD Physics. Aerospace Sciences Meetings. American Institute of Aeronautics and Astronautics, August 2004. DOI: 10.2514/6.2004-412.

Alexander L. Yarin. *Free liquid jets and films*. Interaction of Mechanics and Mathematics Series. Longman Scientific & Technical, Harlow, Essex, 1993. ISBN 0582102952.

Alexander L. Yarin. Drop impact dynamics: splashing, spreading, receding, bouncing... *Annual Review of Fluid Mechanics*, 38:159–192, 2006.

Alexander L. Yarin and Daniel A. Weiss. Impact of drops on solid surfaces: self-similar capillary waves, and splashing as a new type of kinematic discontinuity. *Journal of Fluid Mechanics*, 283:141–173, 1995.

Peng Yi and Gregory C. Rutledge. Molecular origins of homogeneous crystal nucleation. *Annual Review of Chemical and Biomolecular Engineering*, 3:157–182, 2012.

Nomenclature

Latin symbols

A	Surface, m^2
A_c	Accumulation parameter
AH	Absolute humidity, $kg\,m^{-3}$
AoA	Angle of attack
a	Speed of sound, $m\,s^{-1}$
$\vec{a}$	Acceleration vector, $m\,s^{-2}$
C	Constant
CS	Control surface, m^2
CV	Control volume, m^3
c	Chord length, m
c_D	Drag coefficient
c_f	Skin-friction coefficient
c_L	Lift coefficient
c_p	Constant-pressure specific heat, $J\,kg^{-1}\,K^{-1}$
c_v	Constant-volume specific heat, $J\,kg^{-1}\,K^{-1}$
c_0	Speed of light in vacuum, $c_0 = 299\,792\,458\,m\,s^{-1}$
D	Diameter of a primary droplet before impact, m
$\vec{D}$	Dissipative operator, mixed units
$\mathbf{D}$	Diagonal matrix
d	Diameter of an individual entity – e.g. droplet or pore – within a collective, µm
d_{nose}	Nose radius of an aerofoil, m
E	Total energy, $m^2\,s^{-2}$
$\vec{E}$	Electric vector, $V\,m^{-1}$
e	Euler's number, $e = 2.718$
e	Specific internal energy, $m^2\,s^{-2}$
$F, \vec{F}$	Force, N
$\mathbf{F}$	Flux tensor, mixed units
f'	Focal length, m
f	Freezing fraction
f_0	Stagnation-line freezing fraction
$\vec{f}_\sigma$	Volumetric force at an interface, $N\,m^{-3}$
G	Free energy, J
G_v	Driving force of solidification, $J\,m^{-3}$
$\vec{g}$	Vector of gravitational acceleration, $m\,s^{-2}$
H	Total enthalpy, $m^2\,s^{-2}$
h	Height, m
h	Specific enthalpy, $m^2\,s^{-2}$
$\mathbf{I}$	Identify matrix
i	Integer locating the spatial position in a discretised space
K	Turbulent kinetic energy, $m^2\,s^{-2}$
k_B	Boltzmann constant, $1.3807 \times 10^{-23}\,J\,K^{-1}$
k	Thermal conductivity coefficient, $W\,m^{-1}\,K^{-1}$
$\hat{k}$	Complex wave number, m^{-1}
$\mathbf{L}$	Lower triangular matrix
L_{fus}	Latent heat of fusion, $J\,kg^{-1}$
L_{evap}	Latent heat of evaporation, $J\,kg^{-1}$
L_{subl}	Latent heat of sublimation, $J\,kg^{-1}$
Le	Lewis number
LWC	Liquid water content, gm^{-3}
l	Length, m
M	Molecular mass, Da
Ma	Mach number

MVD	Median volume diameter, µm
m	Mass, kg
N	Number
Nu	Nusselt number
n	Refractive index
$\hat{n}$	Complex refractive index
$\vec{n}$	Normal vector
P	Power, W
Pr	Prandtl number
p	Static pressure, Pa
Q	Heat, J
$\dot{q}$	Heat flux, $\mathrm{W\,m^{-2}}$
R	Radius, m
$\vec{R}$	Residual, mixed units
r	Contact radius during droplet impact, m
Re	Reynolds number
RH	Relative humidity
S	Surface, $\mathrm{m^2}$
Sc	Schmidt number
St	Stokes number
Stn	Stanton number
s	Coordinate along an aerofoil surface, m
T	Temperature, K
Tu	Turbulence level
t	Time, s
t_{acc}	Accumulation time of the icing process, s
$\boldsymbol{U}$	Upper triangular matrix
U_∞	Free-stream velocity, $\mathrm{m\,s^{-1}}$
u, v, w	Velocity components in x, y, z-direction, $\mathrm{m\,s^{-1}}$
$\vec{u}$	Velocity vector, $\mathrm{m\,s^{-1}}$
u_τ	Shear velocity, $\mathrm{m\,s^{-1}}$
V	Volume, $\mathrm{m^3}$
W	Number of microstates
$\vec{W}$	Vector of the conservative variables, mixed units
We	Weber number
x, y, z	Coordinates in Cartesian space, m
Y	Similarity parameters for scaling of ice accretion
z'	Camera coordinate, m

Greek symbols

β	Local catch efficiency
γ_{el}	Electrical conductivity, $\mathrm{S\,m^{-1}}$
Δ	Difference or change
δ	Boundary layer thickness, m
$\epsilon^{(2)}, \epsilon^{(4)}$	Indicator functions for artificial dissipation
ε_{el}	Dielectric constant, $\mathrm{N/V^2}$
ε_{rel}	Relative permittivity
Θ	Angle between two crossing laser beams
θ	Contact angle
κ	Von Kármán constant
$\varkappa_{rel}$	Over-relaxation parameter in LU decomposition
Λ	Largest eigenvalue
λ	Wave length, m
μ	Dynamic viscosity, Pa s
μ_{mag}	Magnetic permeability, $\mathrm{N/A^2}$
μ_{rel}	Relative permeability
ν	Kinematic viscosity, $\mathrm{m^2\,s^{-1}}$
$\check{\nu}_T$	Modified kinematic eddy viscosity, $\mathrm{m^2\,s^{-1}}$
ϱ	Density, $\mathrm{kgm^{-3}}$
ϱ_n	Molecule number density, $\mathrm{kgm^{-3}}$
σ	Surface tension, $\mathrm{N\,m^{-1}}$
$\boldsymbol{\tau}$	Viscous shear tensor, $\mathrm{N\,m^{-2}}$
τ_w	Wall shear stress, $\mathrm{N\,m^{-2}}$
Φ	Pivot angle of receiver optics
ϕ	Representative flow variable
φ, φ'	Geometric angle
χ	Scalar indicator of the Volume of Fluid method
Ψ	Elevation of receiver optics
ω	Angular frequency, Hz

Calligraphic symbols

$\mathscr{B}$	Body
$\mathscr{d}$	Distance, m
$\mathscr{f}$	Function
$\mathscr{h}$	Convective heat transfer coefficient, $\mathrm{W\,m^{-2}\,K^{-1}}$
$\mathscr{i}$	Imaginary unit, $\mathscr{i} = \sqrt{-1}$
$\mathscr{k}$	Roughness, µm
$\mathscr{L}$	Van Leer limiter
$\mathscr{O}$	Order of magnitude
$\mathscr{P}$	Point
$\mathscr{R}$	Specifitc gas constant, $\mathrm{J\,kg^{-1}\,K^{-1}}$
$\mathscr{S}$	Entropy, $\mathrm{J\,K^{-1}}$
$\mathscr{w}_{\vec{E}}$	Energy density, $\mathrm{J\,m^{-3}}$
$\odot$	Sphericity

Gothic symbols

$\mathfrak{A}_{\text{bead}}$	Empirical beading factor
$\mathfrak{b}$	Multi-step weight
$\mathfrak{C}$	Contraction of the wind-tunnel nozzle
$\mathfrak{D}$	Non-dimensional diameter
$\mathfrak{d}$	Non-dimensional parameter for droplet impact characterising capillary vs. lubrication pressure
$\mathfrak{H}$	Non-dimensional height
$\mathfrak{h}$	Non-dimensional height
$\mathfrak{K}$	Non-dimensional inertia parameter of a droplet
$\mathfrak{n}$	Ratio of refractive indices
$\mathfrak{P}$	Power factor
$\mathfrak{r}$	Argument of a Van Leer limiter
$\mathfrak{T}$	Non-dimensional thickness
$\mathfrak{t}$	Non-dimensional time
$\mathfrak{w}$	Wetting factor
$\mathfrak{X}$	Local stretching of a computational grid

Sub- or superscripts

air	Air
air gap	Air gap between droplet and wall during dry impact
avail	Available
b	Beam
bead	Bead, an ice roughness element
CS	Control surface
CV	Control volume
cond	Conductive
conv	Convective
cro	Crown of a splashing droplet
D	Drag
d	Droplet
den	Dendrites
e	Edge
eff	Effective
eq	Equivalent
evap	Evaporative
f	Film condition of the flywheel apparatus
face	Face of a polyhedral control volume
fan	Fan
film	Film
fing	Finger
fringe	Fringe
fus	Fusion
G	Glass transition
gas	Gaseous

hole	Hole during crown disintegration
homo	Homogeneous nucleation
hetero	Heterogeneous nucleation
I	Interface
i	Number of a control volume
ice	Ice
imag	Imaginary part
imp	Impact
inlet	Inlet
intf	Interface
j	Vertex number
jet	Jet
kin	Kinetic
lam	Laminar
liq	Liquid
m	Melting
$\mathscr{m}$	Moving
max	Maximum
n	Integer indicating dicretized time
net	Net
nominal	Nominal
numerical	Numerical
P	Particle
PDI	Phase Doppler Interferometry
pore	Pore
prev	Previous
r	Radial
$\mathscr{r}$	Rough wall
rb	Runback
rb,in	Inflowing runback
rb,out	Outflowing runback
real	Real part of a complex number
rec	Recovery
ref	Reference
rel	Relative
res	Residual
rim	Rim of a liquid sheet
S	Surface
s	Sand
$\mathscr{s}$	Smooth wall
sec	Secondary droplets
sens	Sensible
solid	Solidification
splash	Splashing
st	Standing water
T	Turbulent
target	Target of the flywheel apparatus
thBL	Thermal boundary layer
ts	Test section
turb	Turbulent
vap	Vapour
visc	Viscous
w	Water
ws	Water on a model surface
ww	Gaseous water over a layer of liquid water during evaporation
σ	Surface tension
0	Initial
$*$	Critical condition
$+$	Non-dimensionalisation in wall coordinates
$\diamond$	Non-dimensionalization
$\bullet$	Equilibrium condition
$\triangleleft$	Location within the logarithmic overlap region
$\star, \star\star$	Predictor / corrector steps in SIMPLE and PISO
$'$	Fluctuation part of a Reynolds decomposition
$''$	Fluctuation part of a Favre decomposition
∞	Free-stream conditions

Operators

$\mathrm{d_e}, \mathrm{d_i}$	Exchange/internal total differential
$(\cdots)^{\mathrm{T}}$	Matrix transposition
$\langle \phi \rangle$	Ensemble mean
$\overline{\phi}$	Time, or Reynolds average
$\widetilde{\phi}$	Favre average
$\dot{\phi}$	Time derivative
rot	Curl

www.ingramcontent.com/pod-product-compliance
Ingram Content Group UK Ltd.
Pitfield, Milton Keynes, MK11 3LW, UK
UKHW061828190726
13853UKWH00009B/2494

9 783736 972247